ARBITRATION
LAW AND PRACTICE

ARBITRATION LAW AND PRACTICE

PETER M. B. ROWLAND

B.A., LL.B., F.C.A., F.C.I.Arb.

of Gray's Inn and the Inner Temple

Barrister-at-Law

Foreword by the

Rt Hon Lord Wilberforce

The Institute of Chartered Accountants In England and Wales
in association with Sweet & Maxwell Ltd.

In publishing this book the Council of the Institute of Chartered Accountants in England and Wales considers that it is a worthwhile contribution to discussion, without necessarily sharing the views expressed, which are those of the author.

No responsibility for loss occasioned to any person acting or refraining from action as a result of any material in this publication can be accepted by the author or the publishers.

Copyright © 1988 P. M. B. Rowland
ISBN 0 85291 888 7

First published in Great Britain in 1988 by the Institute of Chartered Accountants in England and Wales, Chartered Accountants' Hall, Moorgate Place, London, EC2P 2BJ.

All rights reserved. No part of this publication may be reproduced, stored in a retrieval system, or transmitted in any form or by any means, electronic, mechanical, photocopying, recording or otherwise without the prior permission of the publisher.

Printed and bound by Hobbs the Printers of Southampton

To my wife Clare whose artistic integrity and determination to surmount difficulties have been a constant inspiration

Foreword

It is almost 100 years since the Arbitration Act 1889 was passed. In that time arbitration has moved from being a very junior partner to litigation, to be carefully kept under control by the Courts, to a position of equality, and even dominance, in some important dispute areas. They are now few people in the professions or industry who do not, at some time, come in contact with arbitration.

Naturally, books on arbitration now abound, with a tendancy to increase in volume. Mr Rowland, while responding to the demand, has, happily and skilfully, been able to resist the tendency. He gives us in this book a professional statement of both law and practice, together with some useful precedents, all of this compactly presented in little more than 120 pages. This book is likely to be useful to professional advisers, to non-specialist lawyers facing an arbitration problem, and even to the specialist as a convenient reference link which is wholly up-to-date. I am glad to commend it as a useful addition to libraries.

Richard Wilberforce
September 1987

Contents

	Page
Foreword	vii
Contents	ix
Introduction	xxv
Chapter 1: The nature and scope of arbitration	1
1. Definitions and legislative background	1
2. Arbitration, conciliation and 'ADR'	1
3. Arbitration and valuation	3
4. What may be referred	4
5. Capacity to refer	5
6. Who may act as arbitrator	5
7. The Limitation Acts	7
8. Relationship between litigation and arbitration	8
9. The applicable law(s)	9
10. Equity clauses	10
11. Statutory arbitrations	12
Chapter 2: Advantages and disadvantages of arbitration	14
Arbitration generally preferable	14
1. Cost	14
2. Lack of publicity	15
3. Speed	16
4. Technical expertise	17
5. The enforceability of an award	18
6. Representation	18
7. Extent of jurisdiction	19
8. Convenience	19
9. Flexibility of procedure	19
10. Relationship between tribunal and disputants	20
Litigation generally preferable	20
11. Multi-party disputes	20
12. Strength of the case	22
13. Reference to the European Court of Justice	22

Which is better depends on the circumstances   23
14. Nature of the dispute whether of fact or law   23
15. Legal aid   23

Chapter 3: The arbitration agreement   25
1. Its form and contents   25
2. Implied terms   27
3. Enforcing an arbitration agreement   28
4. Resisting an arbitration agreement   29
5. Applications for a stay   30
   (a) Statutory provisions   30
   (b) Other considerations   31
   (c) Steps in the proceedings   32
   (d) Sufficient reasons not to refer   32
   (e) Existence of a dispute   33
6. Arbitration as a condition precedent to litigation – *Scott* v *Avery* clauses   34
7. Time limitations – *Atlantic Shipping* or *Centrocon* clauses   34
8. Termination of the arbitration agreement   36

Chapter 4: The offices of arbitrator and umpire   38
1. Appointment by the parties   38
2. Umpires   39
3. The number of arbitrators   40
4. Appointment by the Court   41
5. Judicial appointments   42
6. Powers exercisable in an arbitration   43
7. Remuneration   45
8. Removal of an arbitrator or umpire   46
9. Revocation of the authority of an arbitrator or umpire   47
10. Retirement and resignation   48

Chapter 5: Procedure before the hearing   50
1. Initially   50
2. The preliminary meeting   50
   (a) Procedural rules and time-table   51
   (b) Claimant and respondent   52
   (c) Clarifying the issues   53
   (d) Discovery and inspection of documents   54
   (e) Agreed bundles   55
   (f) Inspection and protection of property   55
   (g) Other matters requiring decision   56
   (h) Reasons and exclusion agreements   56
   (i) Preparations for the hearing   57
   (j) Directions   57
3. Later procedural matters   58

4. Pre-trial procedure in London maritime cases	59
5. Termination of the arbitration prior to the hearing	60

Chapter 6: The hearing, alternatives and special cases   61
1. The arbitrator's control of procedure   61
2. High Court procedure   62
3. Some arbitral modifications   63
4. Special procedural considerations   63
5. Attendance at the hearing   65
6. Evidence   65
7. Legal assessors   66
8. 'Documents-only' arbitrations   66
9. 'Look-sniff' arbitrations   67
10. 'String' arbitrations   68
11. 'Pendulum' arbitrations   68
12. Commodity arbitrations generally   68
13. Construction industry arbitrations   69
14. High technology disputes   70

Chapter 7: The Award   72
1. General   72
2. Finality of an award   73
3. Interim awards   74
4. Reasoned awards   75
5. Form   76
6. Currency   78
7. Timing   78
8. Costs   79
   (a) Generally   79
   (b) Taxation   80
   (c) The arbitrator's discretion   82
   (d) Offers prior to the arbitrator's determination   83
   (e) Calderbank offers   84
   (f) VAT upon costs   84
9. Interest   85
10. The arbitrator's lien   86
11. Enforcement   87
   (a) Under S26 of the 1950 Act (as amended)   87
   (b) Enforcement by action   88
   (c) Suing for a declaration or for specific performance   88

Chapter 8: Appeals to, and control by, the Courts   90
1. The Commercial Court   90
2. Appeals under the arbitration agreement   91
3. Appeals under the Arbitration Act 1979   92
   (a) A historical note   92

(b) Determination of a preliminary point of law	93
(c) Appeals – reasons	93
(d) Appeal procedure	94
(e) Appeals – the judicial review of awards	95
(f) Appeals from the High Court	95
(g) The granting of leave to appeal – the *Nema* 'guide-lines'	96
(h) Later modifications to the guide-lines	98
4. Exclusion agreements	99
5. Interventions under the 1950 Act	100
6. The courts' 'inherent jurisdiction' to supervise arbitrations	102

Chapter 9: International arbitrations — 104

1. General	104
2. The arbitration agreement	104
3. Applicable rules and administered arbitrations	106
4. Evidence	107
5. Enforcement of foreign awards	108
Convention awards	108
Awards under the Geneva Convention	110

Appendices

Appendix 1

(a) Directions	112
(b) Scott Schedule	114
(c) Peremptory Notice	115
(d) Award	116

Appendix 2: Legislation and Orders

1. Arbitration Act 1950	121
2. Arbitration Act 1975	144
3. Arbitration Act 1979	148
4. Civil Evidence Act 1968	155
Arbitration (Commodity Contracts) Order (SI 1979 No. 754)	170
Rules of the Supreme Court, Order 73	172
Arbitration (Foreign Awards) Order 1984 (SI 1984/1168 as amended by SI 1985/455, SI 1986/949 and SI 1987/1029)	179
Index	181

Table of Cases

The numbers in bold refer to pages within this book.

Abu Dhabi Gas Liquefaction Co. Ltd. v *Eastern Bechtel Corp.* [1982] 2 Ll. Rep. 425 **21, 48, 61**
Adams v *Catley* (1892) 40 WR 570; 46 SJ 52 **32**
Aden Refinery Co. Ltd. v *Ugland Management Co.* [1986] 3 All ER 737 **96, 98**
Agios Lazaros, The [1976] 2 Lloyd's Rep. 47 **7**
The Agrabele [1979] 2 Ll. R. 117 **31**
Agromet Motoimport Ltd. v *Maulden Engineering Co. (Beds) Ltd.* [1985] 2 All ER 436 **8, 28**
Allied Marine Transport Ltd. v *Vale do Rio Doce Navegacao SA The Leonidas D* [1985] 1 WLR 925; [1985] 2 All ER 796 **29**
Alfa Nord, The – see *Gunnstein* etc.
Andre et Compagnie SA v *Marine Transocean Ltd., The Splendid Sun.* [1981] 1 QB 694; [1981] 125 SJ 395; [1981] 2 All ER 993; [1981] Com. LR 95; [1981] 2 Ll. Rep. 29 **29**
Antaios Cia v *Salen Rederierna* [1984] 3 All ER 229 **76, 94, 97, 98**
Arab African Energy Corp. v *Olieprodukten Nederland BV* [1983] 2 Ll. Rep. 419 **99**
Arenson v *Arenson and* v *Casson, Beckman, Rutley & Co.* [1977] AC 405; 119 SJ 810; [1975] 3 All ER 901; [1976] 1 Ll. Rep. 179 **3**
Aspen Trader, The – see *Libra* etc.
Associated Bulk Carriers Ltd. v *Koch Shipping Inc. The Fuohsan Maru* [1978] 1 Ll. Rep. 24; [1978] 2 All ER 254 **31, 33**
Astro Vencedor Compania Naviera SA v *Mabanaft GmbH, The Damianos* [1971] 2 QB 588, 115 SJ 284 **26**
Atlantic Shipping and Trading Co. v *Dreyfus* [1922] 2 AC 250 **28, 34**
Babanaft International Co. SA v *Avanti Petroleum Inc.*, [1982] 1 WLR 871; [1982] 2 Ll. Rep. 99 **93, 98**
Bangladesh Ministry of Food v *Bengal Liners Ltd.* [1986] 1 Ll. Rep. 167 **39**
Bani and Havbulk v *Korea Shipbuilding and Engineering Corp.* FT Law Reports 10th July, 1987 **10, 46**
Bellshill & Mossend Co-operative Society v *Dalziel Co-operative Society* [1960] AC 832; [1960] 1 All ER 673 **88**

Berkshire Senior Citizens Housing Association v *McCarthy E. Fitt Ltd* (1979) 15 BLR 27 **32**
Blackwell v *Derby Corporation* (1911) 75 JP 129 **32**
Bloemen Property Ltd. v *Gold Coast City*, [1973] AC 115 **89**
Borthwick (Thomas) (Glasgow) Ltd. v *Faure Fairclough Ltd.* [1968] 1 Ll. Rep. 16 **47**
Bremer Handelsgesellschaft mbH v *Ets. Soules* [1985] 2 Ll. Rep. 199 **6**
Bremer Handels GmbH v *Westzucher GmbH* [1981] 2 Ll. Rep. 130 **76**
Bremer Vulkan Schiffbau und Maschinenfabrik v *South India Shipping Corp. Ltd.* [1981] AC 909; [1981] 2 All ER 289; [1981] 1 Ll. Rep. 253 **29, 58, 61, 102**
Brighton Marine Palace Ltd. v *Woodhouse* [1893] 2 Ch. 486; 62 LJ Ch. 697 **32**
Brown v *Llandovery Terra Cotta Co. Ltd.* (1909) 25 TLR 625 **45**
Brown (Christopher) v *Genossenschaft Oesterreichischer Waldbesitzer GmbH* [1954] 1 QB 8; [1953] 2 All ER 1039; [1953] 2 Ll. Rep. 373 **4, 72, 77, 88**
Bulk Oil (Zug) AG v *Sun International Ltd.* [1984] 1 All ER 386 **22, 98**
Bulk Oil (Zug) AG v *Trans-Asiatic Oil Ltd. SA* [1973] 1 Ll. Rep. 129 **31**
Bunge SA v *Kruse* [1979] 1 Ll. Rep. 279; affirmed [1980] 2 Ll. Rep. 142 **37, 38**
Carlisle Place Investments Ltd. v *Wimpey Construction (UK) Ltd.* (1980) 15 BLR 109 **61**
Ceylon (Government of) v *Chandris (No. 1)* [1963] 2 QB 327; [1963] 2 All ER 1; [1963] 1 Ll. Rep. 214 **78, 81**
Chilton v *Saga Holdings* [1986] 1 All ER 841 **61, 62**
Chiswell Shipping Ltd. v *State Bank of India (No. 2)* [1987] 1 Ll. Rep. 157 **73**
Coastal States Trading (UK) Ltd. v *Mebro Mineraloelhandelsgesellschaft mbH* [1986] 1 Ll. Rep. 464 **85**
Cohl (Samuel J.) & Co. v *Eastern Mediterranean Maritime Ltd.* [1980] 1 Ll. Rep. 371 **102**
Compagnie D'Armement Maritime SA v *Compagnie Tunisienne de Navigation SA* [1971] AC 572; [1970] 3 All ER 71 **9**
Conway v *Rimmer* [1968] AC 910 **54**
Courtney & Fairbairn Ltd. v *Tolaini Brothers (Hotels) Ltd.* [1975] 1 All ER 716 **19**
Cutts v *Head* [1984] Ch. 290 **84**
Damianos, The – See *Astro* etc.
Davies, Graham H., (UK) Ltd. v *Marc Rich & Co. Ltd.* [1985] 2 Ll. Rep. 423 **35**
Deutsche Schachtbau-und Teifbohrgesellschaft mbH v *Ras Al Khaimah National Oil Co.* [1987] 2 All ER 769 **11**
Doleman v *Ossett* [1912] 3 KB 257 **8**
Eads v *Williams* (1854) 24 LJ Ch. 531; 24 LT (o.s.) 162; 3 WR 98; 43 ER 671; 4 De G.M. & G. 674; 1 Jur. (NS) 193 **89**

Eagle Star Insurance Co. v *Yuval Insurance Co.* [1978] 1 Ll. Rep. 357 **11, 28, 32**
Eastern Saga, The – see *Oxford* etc.
Edwards v *Bairstow* [1956] AC 14 **12**
Elizabeth H., The [1962] 1 Ll. Rep. 172 **32**
Ellis Mechanical Services v *Wates Construction Ltd.* [1978] 1 Ll. Rep. 33 **31**
Erich Schroeder, The [1974] 1 Ll. Rep. 192 **82**
Eregli, The M. – see *Tradax* etc.
European Grain and Shipping Ltd. v *Dansk Landbrugs Grovvareslskab* [1986] 1 Ll. Rep. 163 **36**
European Grain and Shipping Ltd. v *Johnston* [1982] 3 All ER 989; [1982] Com. LR 12 **78**
Excomm Ltd. v *Ahmed Abdul-Qawi Bamaodah* [1985] 1 Ll. Rep. 403 **26**
Fakes v *Taylor Woodrow Construction Ltd.* [1973] 1 QB 436; [1973] 1 All ER 670; 117 SJ 13 **33**
Folias, The – see *Services* etc.
Food Corporation of India v *Antclizo* [1987] 2 Ll. Rep. 130 **29**
Food Corporation of India v *Marastro Cia Naviera SA The Trade Fortitude* [1986] 3 All ER 500; [1986] 2 Ll. Rep. 209 **73**
Frankenburg and the Security Co. (1894) 10 TLR 393 **7**
French Government v *Tsurushima Maru SS.* (1921) 37 TLR 961; 8 Ll. Rep. 403 **10**
Fuohsan Maru, The – see *Associated* etc.
Getreide-Import GmbH v *Contimar SA Compania Industrial Commercial y Maritima* [1953] 1 WLR 793; 97 SJ 434; [1953] 2 All ER 223; [1953] 1 Ll. Rep. 572 **4**
Goodman v *Winchester and Alton Railway* [1984] 3 All ER 594 **33**
Gray v *Ashburton (Lord)* (1917) AC 26; 86 LJKB 224; 115 LT 739; 81 JP 17 **83**
Gunnstein & Co. v *Jensen Krebs and Nielsen, The Alfa Nord* [1977] 2 Ll. Rep. 434 **33**
Haigh v *Haigh* (1861) 31 LJ Ch. 420; 3 De G.F. & J. 157; 8 Jr. (NS) 983; 5 LT 507 **65**
Hannah Blumenthal, The – see *Paal* etc.
Heaven and Kesterton v *Sven Widaeus A/B* [1958] 1 Ll. Rep. 101; [1958] 1 WLR 248 **83**
Heyman v *Darwins Ltd.* [1942] AC 356; 111 LJKB 241; 166 LT 306 58 TLR 169; [1942] 1 All ER 337 **36**
Higgs and Hill Building Ltd. v *University of London* (1983) 24 BLR 139 **70, 83, 97**
Himmerland, The [1965] 2 Ll. Rep. 353 **36**
Hobbs Padgett & Co. (Reinsurance) Ltd. v *J. C. Kirkland Ltd.* 113 SJ 832; [1969] 2 Ll. Rep. 547 **25**
India (President of) v *La Pintada Cia Navegacion SA* [1985] AC 104; [1984] 2 All ER 773 **85**

India (President of) v *Lips Maritime Corp.* [1987] 3 All ER 110 **85**
Interbulk Ltd. v *Aiden Shipping Co. Ltd. The Vimeira* [1984] 2 Ll. Rep. 66 **21**
International Sea Tankers Inc. v *Hemisphere Shipping Co. Ltd. The Wenjiang* [1982] 1 Ll. Rep. 128 **97**
Italmare Shipping Co. v *Ocean Tanker Co. Inc. The Rio Sun* [1981] 2 Ll. Rep. 489 **97**
Jackson v *Barry Railway Co.* [1893] 1 Ch. 238; 68 LT 472; 9 TLR 90 **7**
Jamieson and Binns, Re (1836) 4 A & E 945; 5 LJ (NS) KB 187; 43 RR 527; 111 ER 1039 **39**
Jedranska Slobodna Plovidba v *Oleagine SA The Luka Botic* [1983] 3 All ER 602 **35**
Knowles and Sons Ltd. v *Bolton Corporation* [1900] 2 QB 253; 69 LJ QB 481; 82 LT 229; 16 TLR 283; 48 WR 433 **79**
Kostas Milas, The – see *Sethia* etc.
Kuwait Ministry of Public Works v *Sir Frederick Snow* [1984] 1 All ER 733 **109**
Laertis Shipping Corp. v *Exportadora Española de Cementos Portland SA The Laertis* [1982] Com. LR 298; [1982] 1 Ll. Rep. 613 **40-1**
Leonidas D. The – see *Allied* etc.
Liberian Shipping Corporation v *King (A.) & Sons Ltd.* [1967] 2 QB 86; [1967] 2 WLR 856; 111 SJ 91; [1967] 1 All ER 934; [1967] 1 Ll. Rep. 303 **35**
Libra Shipping and Trading Corporation Ltd. v *Northern Sales Ltd. The Aspen Trader* [1981] 1 Ll. Rep. 273 **35**
Lloyd v *Wright* [1983] 2 All ER 969 **8**
London, Dover and Chatham Rly v *South Eastern Rly.* [1893] AC 429 **85**
Lucas Industries Plc. v *Welsh Development Agency* [1986] 1 Ch. 550 [1986] 2 All ER 858 **98**
Luka Botic, The – see *Jedranska* etc.
Mafracht v *Parnes Shipping Co. SA The Apollonius* [1986] 2 Ll. Rep. 405 **72**
Mariana Islands Steamship Corp. v *Marimpex Mineraloelhandelsgessellschaft mbH The Medusa* [1986] 2 Ll. Rep. 328 **35**
Medusa, The – see *Mariana* etc.
Miller (James) & Partners v *Whitworth Street Estates (Manchester) Ltd.* [1970] AC 583; 114 SJ 225; [1970] 1 All ER 796; [1970] 1 Ll. Rep. 269 **10**
Mitsubishi v *Soler Chrysler-Plymouth* (1985) 87 L. Ed. 2d. 444 **111**
Montana, The – see *Mutual* etc.
Mondial Trading Co. GmbH v *Gill and Duffus Zuckerhandels GmbH* [1980] 2 Ll. Rep. 376; (1979) 15 Build. LR 118 **95**
Monro v *Bognor UDC* [1915] 3 KB 167; 84 LJKB 1091; 112 LT 969; 59 SJ 348 **26**
Moran v *Lloyd's* [1983] 2 All ER 200 **101**

Mutual Shipping Corp. v *Bayshore Shipping Co. The Montana* [1985] 1 Ll. Rep. 189, [1985] 1 All ER 520 **73, 74, 76**
National Rumour Co. SA v *Lloyd-Libra Navegacao SA* [1982] 1 Ll. Rep. 472; [1982] Com. LR 4 **97**
National Westminster Bank Plc v *Arthur Young McLelland Moores & Co.* [1985] 2 All ER 817 **96**
Naviera Amazonia v *Compania Internacional de Peru*, The Independent 11 November 1987 **9**
Neale v *Ledger* (1812) 16 East 51; 14 RR 283; 104 ER 1008 **39**
Neale v *Richardson* [1938] 1 All ER 753 **24**
Nema, The – see *Pioneer Shipping* v *B. T. P. Tioxide*
Norske Atlas Insurance Co. Ltd. v *London General Insurance Co. Ltd.* (1927) 43 TLR 541 **88**
Northern Regional Health Authority v *Derek Crouch Construction Co. Ltd.* [1984] 2 All ER 175 **9, 11, 16, 19, 72**
Nova (Jersey) Knit v *Kammgarn Spinnerei GmbH* [1977] 1 WLR 713; 121 SJ 170; [1977] 2 All ER 463; [1977] 1 Ll. Rep. 463 **31, 33**
Nuttall v *Manchester Corporation* (1892) 8 TLR 513 **7**
Oakland Metal Co. Ltd. v *Benaim (D) & Co.* [1953] 2 QB 261; 97 SJ 540; [1953] 2 All ER 650; [1953] 2 Ll. Rep. 192 **79**
Orion Compania Espanõla de Seguros v *Belfort Maatschappij Voor Algemene Verzehgringeen* [1962] 2 Ll. Rep. 257 **10, 11, 12**
Overseas Fortune Shipping Pte. Ltd. v *Great Eastern Shipping Co. Ltd.* [1987] 1 Ll. Rep. 271 **100**
Oxford Shipping Co. Ltd. v *Nippon Yusen Kaisha, The Eastern Saga* [1984] 3 All ER 835, [1984] 2 Ll. Rep. 373 **21, 28**
Paal Wilson & Co. A/S v *Partenreederei Hannah Blumenthal (The Hannah Blumenthal)* [1983] 1 AC 854; [1982] 1 Ll. Rep. 582; [1982] Com. LR 117; 126 SJ 292 [1983] 1 All ER 34 **28, 30, 37**
Perkins (H. G.) Ltd. v *Brent-Shaw* [1973] 1 WLR 975 **81**
Phonizien, The [1966] 1 Ll. Rep. 150 **29**
Pioneer Shipping v *B. T. P. Tioxide, The Nema* [1982] AC 724; 125, SJ 542; [1981] 2 All ER 1030; [1981] 2 Ll. Rep. 239; [1981] Com. LR 197 **15, 40, 93, 94, 95, 96–9**
Pitchers Ltd. v *Plaza (Queensbury) Ltd.* [1940] 1 All ER 151; 162 LT 213; 56 TLR 257; 84 SJ 76 **32**
Pittalis v *Sherefettin* [1986] QB 868; [1986] 2 All ER 227 **5, 27, 35**
Practice Notes:
  [1985] 2 All ER 383 **94**
  [1985] 2 All ER 384 **103**
  [1987] 3 All ER 799 **17**
Printing Machinery Co. Ltd. v *Linotype and Machinery Ltd.* [1912] 1 Ch. 566; 81 LJ Ch. 422; 106 LT 743; 28 TLR 224 **26**
Rahcassi Shipping Co. SA v *Blue Star Line Ltd.* [1969] 1 QB 173; [1967] 3 WLR 1382; 111 SJ 848; [1967] 3 All ER 301; [1967] 2 Ll. Rep. 261 **6**

Richards v *Payne & Co.* (1916) 86 LJKB 937; 115 LT 225; 13 Asp. MLC 446 **26**
Rio Sun, The – See *Italmare* etc.
Rocco Giuseppe et Figli v *Tradax Export SA* [1983] 3 All ER 598 **86**
Roussel-Uclaf v *Searle (G. D.) & Co.* [1978] 1 Ll. Rep. 225 **32**
Rowcliffe v *Devon and Somerset Railway Co.* (1873) 21 WR 433 **81**
S. L. Sethia – See *Sethia*
Scammell v *Ouston* [1941] AC 251; 110 LJKB 197; 164 LT 379; 57 TLR 379; 85 SJ 224; [1941] 1 All ER 14; 46 Com. Cas. 190 **19**
Scott v *Avery* [1856] HL Cas. 811; 25 LJ Ex. 308; 28 LT (o.s.) 207 **28, 34, 102**
Sellar v *Highland Ry.*, [1919] SC (HL) 19; 56 ScLR 216 **7**
Services Europe Atlantique Sud v *Stockholms Rederiaktiebolag Svea, The Folias* [1979] AC 685 **78**
Sethia Liners Ltd. v *Naviagro Maritime Corp. The Kostas Milas* [1981] 1 Ll. Rep. 18 **33**
Sethia Liners Ltd. v *State Trading Corp. of India Ltd.* [1986] 2 All ER 395 **33**
Shearson American Express v *McMahon* (1987)
Simmons v *Sec. of State for the Environment* [1985] JPL 253 **47**
Smith v *Hartley* (1851) 10 CB 800 **77**
Splendid Sun, The – See *Andre*, etc.
Stainless Patriot, The – See *Whitehall* etc.
Stotesbury v *Turner* [1943] KB 370; 112 LJKB 365; 168 LT 355 **83**
Sutcliffe v *Thackrah* [1974] AC 727; 118 SJ 148; [1974] 1 All ER 859; [1974] 1 Ll. Rep. 318 **3**
Sykes (F & G) (Wessex) Ltd. v *Fine Fare Ltd.* [1967] 1 Ll. Rep. 53 **19**
Takamine, The – See *Wilhelmsen*, etc.
Taunton-Collins v *Cromie* [1964] 1 WLR 633; [1964] 2 All ER 322; 108 SJ 277 **32**
Thorburn v *Barnes* [1867] LR 2 CP 384; 36 LJCP 184; 16 LT 10; 15 WR 623 **26**
Timber Shipping Co. SA v *London and Overseas Freighters Ltd.* [1972] AC 1; [1971] 2 All ER 599 **86**
Toller v *Law Accident Insurance Society Ltd.* [1936] 2 All ER 952; 80 SJ 633 **31**
Tradax Export Internacional SA v *Cerrahogullari TAS The M. Eregli*, [1981] 3 All ER 344; [1981] Com. LR 144; [1981] 2 Ll. Rep. 169 **33, 36**
Trade Fortitude, The – See *Food*, etc.
Tramountana Armadora SA v *Atlantic Shipping Co. SA* [1978] 1 Ll. Rep. 391 **82, 83, 84**
Trave Schiffahrts GmbH & Co. KG v *Ninemia Maritime Corp* [1986] 1 Ll. Rep. 393 [1986] 2 All ER 244 **74, 76, 94**
Turner and Goudy v *McConnell* [1985] 2 All ER 34 **32**
The Tuyuti [1984] 2 All ER 545 **9, 31**

Tzortzis v *Monark Line A/B* [1968] 1 WLR 406; 112 SJ 108; [1968] 1 All ER 949; [1968] 1 Ll. Rep. 337 **41**
Universal Petroleum Co. Ltd. v *Handels-und-Transport GmbH* [1987] 1 Ll. Rep. 517 **94**
Vasso v *Vasso* [1983] 3 All ER 211 **93**
Vimeira – See *Interbulk* etc.
Wadsworth v *Lydall* [1981] 1 WLR 598 **85**
Walford, Baker & Co. v *Macfie* (1915) 84 LJKB 2221; 113 LT 180 **47**
Wallal v *Bank Millat* [1986] 1 All ER 239 **111**
Walters v *Morgan* (1792) 2 Cox Ch. Cas. 369; 30 ER 169 **89**
Warde v *Feedex International Inc.* [1984] 1 Ll. Rep. 310 **75**
Warinco AG v *Andre et Cie SA* [1979] 2 Ll. Rep. 298 **82**
Wenjiang, The – See *International* etc.
Westfal-Larsen & Co. A/S v *Ikerigi Compania Naviera SA* [1983] 1 All ER 382 **27**
Whitehall Shipping Co. v *Kompas Schiffahrtskontor GmbH The Stainless Patriot* [1979] 1 Ll. Rep. 589 **101**
Wilhelmsen v *Canadian Transport Co. The Takamine* [1980] 2 Ll. Rep. 204 **22**
Willcock v *Pickford Removals Ltd.* [1979] 1 Ll. Rep. 244 **4**
Woolf v *Collis Removal Service* [1948] 1 KB 11; [1947] LJR 1377; 177 LT 405; 63 TLR 540; [1947] 2 All ER 260 **5**
Yeates v *Caruth* [1895] 2 IR 146 **39**
Zambia Steel & Buildings Supplies Ltd. v *James Clark and Eaton Ltd.* FT Comm. LR 15 August 1986 **27**
Zermalt Holdings SA v *Nu-Life Upholstery Repairs Ltd.* [1985] 2 EGLR 14 **18, 64, 91, 94**

Table of Statutes

The numbers in bold refer to pages within this book

1677 STATUTE OF FRAUDS (29 Car. 2, c. 3) **27, 89**
1875 PUBLIC HEALTH ACT (38 + 39 Vict. c. 55) **79**
1925 LAW OF PROPERTY ACT (15 & 16 Geo. 5, C. 20) s40 **27**
1950 ARBITRATION ACT (14 Geo VI c 27)
    s1 **47**
    s2
    ss (1) **13, 36**
      (2) **48**
      (3) **36**
    s3 **13**
    s4
      (1) **30**
    s5 **13**
    s6 **27, 40**
    s7 **28, 38, 39, 47**
    s8 **39**
      (1) **27**
      (2) **27**
      (3) **40**
    s9 **40, 78**
    s10 **28, 38, 41, 42**
    s11 **42**
    s12 **43**
      (1) **27, 43, 54, 63, 65**
      (2) **27, 65**
      (3) **43, 65**
      (4) **44**
      (5) **44**
      (6) **44, 45, 90**
        (a) **46, 101**
        (b) **54, 59**
        (e) **56**
        (f) **31**
        (g) **56**

s13
- (1) **79**
- (2) **79**
- (3) **17, 34, 46, 79**

s14 **27, 74**

s15 **27, 73, 89**

s16 **28, 73**

s17 **28, 73**

s18
- (1) **28, 80**
- (2) **81**
- (3) **13, 80**
- (4) **73, 80**
- (5) **80**

s19
- (1) **86**
- (2) **86**

s19A **28, 85**

s20 **85**

s21 (repealed) **92**

s22 **73, 94, 101**
- (1) **86**
- (2) **79**

s23 **101**
- (1) **34, 47**
- (2) **86, 94**
- (3) **73, 87**

s24 **13, 31, 34, 48**
- (1) **7, 47**
- (2) **4, 37, 48, 100**
- (3) **101**

s25 **13, 41, 42**
- (1) **42**
- (2) **34, 37, 42, 102**
- (4) **34, 42**

s26 **11, 18, 87, 89, 109, 111**
- (1) **87**
- (2) **87**
- (3) **87**

s27 **13, 35, 36**

s28 **80**

s29 **13**

s30 **5**

s31 **13, 35**

s32 **25**

s35 **110**
s36 **108, 111**
   (1) **11, 87**
s37 **108**
   (1) **111**
   (2) **111**
   (3) **111**
s38 **108, 111**
1957 HOUSING ACT (5 + 6 Eliz. 2 c. 27) **12**
1962 BUILDING SOCIETIES ACT (10 & 11 Eliz. 2, c. 37) **52**
1968 CIVIL EVIDENCE ACT (c. 64) **155**
1970 ADMINISTRATION OF JUSTICE ACT (c. 31)
   s4 **42, 49**
     (5) **43**
1974 FRIENDLY SOCIETIES ACT (c. 46) **12**
1975 ARBITRATION ACT (c. 3)
   s1
     (1) **30, 31, 33**
     (2) **30**
     (4) **30**
   s3 **108**
   s4 **108**
     55, 108, 109
   s7 (1) **37, 109**
     (2) **109**
   s8 (2) **13, 80**
1977 ADMINISTRATION OF JUSTICE ACT (c. 38) s17 **87**
1979 ARBITRATION ACT (c. 42) **15**
   s1 **76, 91, 92, 93, 99, 101**
     (1) **101**
     (2) **79, 94, 95**
     (3) **22, 93, 95, 96**
     (4) **95, 97**
     (5) **75, 93, 94**
       (b) **94, 96**
     (6) **76, 94**
       (a) **57**
     (6A) **96, 97**
     (7) **96**
       (b) **98**
     (8) **95**
   s2 **66, 93**
     (1) **22, 93**
       (a) **96, 99**
     (2) **93**

  (2A) **96**
  (3) **96**
 s3
  (1) **99**
   (c) **93**
  (2) **99**
  (3) **48, 100**
  (4) **99**
  (5) **13, 99**
  (6) **99**
  (7) **37, 99**
 s4
  (1) **100**
  (3) **100**
 s5 **43, 44, 45**
  (1) **44**
  (2) **44**
 s6 **39**
  (2) **40, 78**
  (3) **42**
  (4) **42**
 s7
  (1)
   (c) **5**
   (d) **13**
   (e) **37**
1980 LIMITATION ACT (c. 58)
 s34 **7-8, 13, 100**
1981 SUPREME COURT ACT (c. 54)
 s148
  (2) **96**
  (3) **96**
1982 ADMINISTRATION OF JUSTICE ACT (c. 53) **85**
1986 AGRICULTURAL HOLDINGS ACT (c. 5) **52**

ORDERS AND RULES
1979 ARBITRATION (COMMODITY CONTRACTS) ORDER (S.I. 1979/754) **100, 170**
1984 ARBITRATION (FOREIGN AWARDS) ORDER (S.I. 1984/1168 as amended by SI 1985/455, 1986/949 and 1987/1029) **109, 110, 179**
1985 JUDGMENT DEBTS (RATE OF INTEREST) ORDER (S.I. 1985/437) **86**

RULES OF THE SUPREME COURT **20**

O.14 **16, 22, 31, 32, 33**
O.18 **53**
O.29 **22**
O.62 **28, 80**
O.73 r.5 **79, 93, 94**
　　　r.10 **87**

COUNTY COURT RULES **10, 20**

Introduction

My aim in this book has been to set out as succinctly as I could the best modern arbitral practice together with the applicable law as at 1 January 1988.

The rapid development which has recently occurred in both law and procedure have not made it an easy task.

So far as the law is concerned, while the most important decisions since the 1979 Arbitration Act, *The Nema* and *Bremer Vulkan*, were in time for the current editions of the major works on the subject – Russell's *Arbitration* (20th Ed.) and Mustill and Boyd's new *Commercial Arbitration* – enough has happened since to justify reference to some 40 later decisions. Producing the present slim volume accordingly involved fairly drastic pruning though I think with little real loss. Changes in the law have been such that many old decisions are misleading except on the very point for which they are cited; besides recent ones are necessarily much richer in the authorities to which they refer.

In regard to practice, I have tried to steer a middle course between the traditionalists who tend to view arbitration as 'wigless litigation' (to use Dr Hermann's graphic phrase) and those who have become impatient of legal formalities or perhaps unduly enamoured of Civil Law procedures, American alternative dispute resolution, or whatever. I can only hope that Rowlatt, J.'s comment (at 12 TC 1004) on a Governmental mix-up does not become apposite: 'There being a space between the two stools, perhaps it is not wholly surprising that the British taxpayer sought repose upon it.' I have however derived reassurance from Bernstein's *Handbook of Arbitration Practice* which was published after most of the text was written since it generally adopts a similar stance.

The extent to which an arbitrator should take the initiative is one source of present day controversy. Some favour the policy of letting sleeping dogs lie, and the reputation of arbitration has suffered as a result. It may, of course, be that the parties have other things on their minds, but the inordinate delays which have occurred in some well-known cases is more suggestive of the ponderous processes of lawyers than foot-dragging by clients.

In general, I take the view that since S12(1) of the Arbitration Act, 1950, indicates, and case law generally confirms, that the arbitrator is master of the proceedings before him, he should not abdicate his authority too readily. After all, if the parties think an arbitrator is going too fast they only have to say so jointly.

Chapter 1 sets out some of the parameters of arbitration as a method of settling disputes consensually and is followed by consideration of its main advantages and disadvantages. Chapter 3 deals briefly but reasonably fully with the legal position in regard to stays and Chapter 8 does the same for appeals. The remainder is mainly devoted to the actual performance of the arbitrator's tasks in domestic arbitrations though international and statutory arbitrations are treated shortly in Chapter 9 and Chapter 1, para. 11. Arbitrations between employers and unions in regard to wages claims, and other arbitrations of a specialised character, are outside the scope of this book.

In Appendix 1 will be found illustrations of some of the documentation which would be appropriate in the circumstances mentioned in the text. Finally, the relevant legislation is set out in full in Appendix 2 together with some extracts from the Rules of the Supreme Court.

I conclude by expressing my warmest thanks and appreciation to Lord Wilberforce who took on the unenviable task of writing a Foreword at a time of great personnel inconvenience; and to John Tackaberry QC, who read through the manuscript, saved me from one or two pitfalls and made many helpful suggestions.

Abbreviations

The following comments on some of the words and phrases used apply unless the context indicates otherwise:

ADR means 'alternative dispute resolution' and is taken here as including conciliation and 'mini-trials' but as excluding arbitration, though in the US it is often used to cover any alternative to litigation and therefore as including arbitration;
The arbitrator means the arbitral tribunal whether it consists of a single arbitrator, as in the great majority of domestic arbitrations, of more than one arbitrator, or an umpire;
The other party means the other party or parties. It has been convenient to assume throughout that there are only two parties as there are seldom more;
The 1950 Act means the Arbitration Act, 1950;
The 1975 Act means the Arbitration Act, 1975;
The 1979 Act means the Arbitration Act, 1979;
The Court means the High Court or a judge thereof;
An **arbitration agreement** means (following s32 of the 1950 Act) a written agreement to submit present or future differences to arbitration, whether an arbitrator is named therein or not;
GAFTA means The Grain and Feed Trade Association;
ICC means the International Chamber of Commerce;
ICE means the Institution of Construction Engineers;
LCIA means the London Court of International Arbitration;
LMAA means London Maritime Arbitrators Association;
Mustill and Boyd means *Commercial Arbitration* by Mustill and Boyd;
RSC means Rules of the Supreme Court;
Russell means Russell's *Arbitration* 20th Edition;
Submission means the actual reference of a dispute to an arbitrator as distinguished from an agreement to refer;
UNCITRAL means the United Nations Commission for International Trade Law.

1 The Nature and Scope of Arbitration

1. Definitions and legislative background

In the sense used in this book 'arbitration' is a consensual system of judicature directed to the resolution of commercial disputes in private.

The term has other meanings; thus, a number of different statutes use it to describe the procedures which they prescribe as alternatives to litigation in the courts. Some consideration to these 'statutory arbitrations' is given at pp.12–13.

Agreements in writing to submit present or future differences to arbitration receive positive encouragement from the legislature in the shape of a miscellany of rules most of which apply unless excluded expressly or by implication. As indicated by the list of Abbreviations, three Arbitration Acts are now in force – the Act of 1950 which provides most of the partial code just mentioned together with some basic provisions which the parties cannot override, that of 1975 which was introduced to give effect to the 1958 New York Convention on the Recognition and Enforcement of Foreign Arbitral Awards, and that of 1979 which is mainly concerned to govern and restrict intervention by the courts.

2. Arbitration, conciliation and 'ADR'

Conciliation, sometimes called mediation is favoured in some countries as a method of resolving disputes, particularly in the Far East. Its aim, that of persuading the parties to reach a settlement by devising terms which both are prepared to accept, is quite different from that of arbitration, which is to determine the dispute which has arisen in accordance so far as may be with their rights.

Usually, where both parties wish to settle, their executives can reach agreement between themselves. However, sometimes the assistance of experienced third parties may be needed to break down prejudices, or assuage feelings of anger or wounded pride – though whether executives who require such industrial psychiatry should be indulged or replaced may be open to doubt in a robust economy.

However that may be, the ICC, UNCITRAL, the LCIA and the Euro-Arab Chambers of Commerce are among the many organisations which have produced rules for conciliators though the procedure suffers from the obvious weakness that a conciliator's views do not bind a reluctant adversary. It follows that if laying one's cards on the table and expending costly executive time fails to induce a reasonable attitude in the other side, the only courses left are to compromise on unsatisfactory terms or resort to litigation or arbitration – perhaps at a disadvantage because of what has transpired notwithstanding the confidential nature of official conciliation procedures.

In the United States of America various other methods of ADR are becoming increasingly popular, the most fruitful manifestation being the mis-named 'mini-trial'.

A 'mini-trial' may take various forms but commonly comprises a short presentation of the issues by the respective in-house lawyers in front of a senior executive(s) from each side who are preferably unconnected with the actual dispute. There may also be a neutral chairman, commonly a lawyer respected by both sides, to elucidate any problems which may arise during the presentation. The executives then retire and try to negotiate a settlement. If they fail, the chairman, if any, may then be asked his view as to the likely result should litigation ensue after which the executives make a further attempt or attempts to settle. If they succeed, the terms are forthwith incorporated into a written argument enforceable under the ordinary law of contract though without assistance from the 1958 Convention – as to which see Chapter 9.

Lapse of a defined period, say, one week, without a settlement will be deemed a failure and litigation or arbitration will presumably follow.

Such procedures are clearly much cheaper than litigation, or even arbitration, provided they work, especially if in-house ('tame') lawyers are used. In addition, they may help to preserve or even cement commercial relationships which have previously subsisted, rather than completing their destruction as so often results from litigation. They are therefore particularly apt in on-going relationships and in cases where there is room for some give and take, but may be a waste of time where the issues are clear-cut.

The claim that ADR tends to induce such reasonableness in the parties that they usually end up settling needs to be considered against the background of a legal system which in many respects is more expensive and less efficient than its English counterpart with the result that almost any alternative to litigation tends to be welcome.

For ADR in relation to 'high-tech' disputes see Chapter 6 para. 14, *post*.

3. Arbitration and valuation

The difference between arbitration and valuation is important because of the number of powers given to the court, and other statutory provisions which only apply in the case of arbitration, and the probability though not certainty that, because of the judicial nature of his functions, an arbitrator cannot be sued for negligence. It may, however, sometimes be difficult to draw the distinction in practice, and using such terms as 'mere valuer', 'quasi-arbitration', and 'arbitration properly so called' makes the task no easier.

While an arbitrator and a valuer may each determine a valuation question mainly on the basis of his own skill and knowledge, it is anticipated that the former will, or at any rate may, finally make up his mind only after embarking on some form of judicial enquiry, however rudimentary, and that he would be in a position to give, if he does not in fact give, a reasoned judgment.

In *Sutcliffe* v *Thackrah* [1974] AC 727, it was held that neither the parties' agreement that certificates of the building owner's architect should be conclusive nor the architect's duty to act fairly between them made him into an arbitrator, and he was liable for interim certificates he wrongly issued.

In *Arenson* v *Arenson* [1977] AC 405, auditors had been expressly instructed to value shares as experts not as arbitrators. This case finally demolished the earlier view that a person whose role is to decide a question which involves him in holding the scales fairly between two parties is immune from an action for negligence. To gain such immunity, if indeed it can be gained at all, the person claiming it must establish that there is a 'sufficient judicial element'. The existence of a dispute is often the vital distinguishing feature of an arbitration since resolving a dispute is the essence of a judicial decision. However, in borderline cases all the circumstances must be taken into account such as whether evidence was called or rival contentions argued, though neither of these is necessary in the case of 'look–sniff' arbitrations. The description given to the tribunal is clearly of considerable importance though appointment as an 'arbitrator' does not of itself necessarily conclude the matter – see the *Arenson* case *op. cit.* at pp. 440 D and 442 D.

It may be added that a valuer who does hear evidence and submissions may sometimes elevate himself into the status of an arbitrator with the parties' consent, and also that the usual method of remunerating valuers, namely, by a percentage of the value ascertained, would be objectionable in the case of an arbitrator by giving him a personal interest in the amount of his award.

4. What may be referred

The general rule is that all matters in dispute concerning personal or real property or a civil wrong may be referred to arbitration. The fact that the dispute is one of law is no longer considered as necessarily rendering it unsuitable although, at any rate before the 1975 Act came into force, it was sometimes treated as a reason for not granting a stay.

However, where there are allegations of fraud the court has power to refuse to enforce a domestic arbitration agreement and may even give leave for revocation of the authority of an arbitrator already appointed – see S24(2) of the 1950 Act. It is more likely to exercise this power at the instance of the party against whom the charge is made.

Nor can anything be referred under a contract which it is against public policy to enforce, or which offends the country's 'most basic notions of morality and justice' (to use US phraseology – see US 517 F. 2 d. at p. 516 2nd Cir. 1975). The appropriate Latin maxim is: *Ex dolo malo non oritur actio.*

An 'arbitrator' cannot give a ruling as to the validity of the contract from which his authority stems since, logically, he is unable to determine whether he has jurisdiction to deal with it – see per Devlin J., in *Christopher Brown Ltd.* v *Genossenschaft Oesterreichischer Waldbesitzer* [1953] 2 All ER 1039, at p. 1042.

For a similar reason, he cannot be asked to rule that there is no dispute, or as to whether antecedent facts necessary to his appointment such as the giving of due notice have occurred – see *Getreide – Import GmbH* v *Contimar SA* [1953] 1 WLR 793 at p. 806, (CA). Again, when a question arose as to whether the contract between the parties included an arbitration agreement the Court of Appeal held that the 'arbitrator' had no jurisdiction to determine it – *Willcock* v *Pickford Removals Ltd.* [1979] 1 Ll. Rep. 244 (CA).

On the other hand, the parties may give the arbitrator express power to determine his jurisdiction – see, for instance B(4) of the Rules of the

Chartered Institute of Arbitrators – and may validly refer to arbitration the specific issue of whether there is jurisdiction in a case of alleged illegality where the illegality does not strike at the arbitration agreement itself. Moreover, an arbitrator may normally determine a question of *subsequent* illegality or as to the *continued* existence of the main contract, and also the extent of his own powers in relation to the conduct of the arbitration.

If an arbitrator's jurisdiction is challenged but there are other matters in dispute between the parties he should hear argument on the jurisdiction point first, not because he has power to rule on the matter – he does not – but in order to decide what course to adopt. If he forms the view that he has no jurisdiction he should proceed no further. If he is in doubt, he should probably not proceed but might do so, *ex parte* if necessary, at the request of one of the parties on receiving an indemnity for his remuneration. It would then be left to the party challenging his jurisdiction to apply for an injunction or to challenge his award in due course.

As mentioned in Chapter 2 para. 7 matters can be referred to an arbitrator which could not be litigated. Thus he may make a bargain for the parties if given the power to do so.

5. Capacity to refer

The Crown may be a party to an arbitration and is bound by both Part 1 of the 1950 Act and by the 1979 Act – see Ss30 and 7(1)(c) of the respective Acts.

An infant's power to refer disputes to arbitration is the same as he has to contract, that is to say, the reference will be voidable by him unless it relates to the supply either of his own services on reasonable terms, or of necessities, or unless it is otherwise for his benefit.

An arbitration agreement may validly entitle only one of the parties to submit disputes to arbitration – see *Woolf* v *Collis Removal Service* [1948] 1 KB 11 at p. 18, and *Pittalis* v *Sherefettin* [1986] 2 All ER 227 (CA), where dicta to the contrary were disapproved.

6. Who may act as arbitrator

In general, any natural person, whatever his age or disabilities, may act as an arbitrator. Differing views have been expressed as to the eligibility of an infant; probably the correct modern view is that he may act subject to the court's discretionary power to intervene in appropriate circumstances.

A person is disqualified if he is a necessary witness in the case, or if he is dishonest or lunatic.

Incompetence is not a disqualification. The general rule is that provided the parties act with their eyes open they may appoint whomever they wish. Each must normally abide by his choice.

The fact that a person is not disqualified in no way predicates his suitability. It is obviously desirable that an appointee should have had experience and some actual training, and it has been said that 'there is no role in modern arbitration practice for arbitrators who lack the courage and initiative to lead' (Shilston).

Indeed, much depends upon the personality of the arbitrator himself and a bad choice can dissipate the advantages which should otherwise obtain. In general, efforts should be made to secure the appointment of an arbitrator who stands no nonsense, is not so busy that he cannot be flexible in fixing dates, and is prepared to give an appropriate degree of priority to the affairs of those who have employed him.

A corporation, other presumably than a corporation sole, cannot be appointed as an arbitrator or umpire. The appointment of a club might possibly be construed as meaning a member(s) of the club duly appointed for the purpose by its governing body, or the governing body itself.

The arbitration agreement may cut down the general capacity of individuals to act by requiring the possession of specified qualities or by stating certain classes of person to be ineligible. Thus, in *Rahcassi Shipping Co. SA v The Blue Star Line Ltd.* [1969] 1 QB 173 the arbitration agreement said that 'arbitrators and umpires shall be commercial men and not lawyers'. The arbitrators, having disagreed, appointed a practicing barrister as umpire and continued as arbitrator-advocates. On challenge to the award, the court held that by appearing before him the arbitrator-advocates had waived the defect in his appointment but as they had no direct authority to do so Roskill, J., had no option but to declare the whole proceedings void.

The appointment of a particular person(s) may be barred as a result of his relationship with the facts or a party, namely, if it is such as would incline a reasonable man to think he would be biassed.

An allegation of bias will not be accepted lightly – see, for instance, *Bremer Handels GmbH v Ets. Soules* [1985] 2 Ll. Rep. 199, where it was doubted whether mere allegations of actual bias would be sufficient to raise a reasonable suspicion of imputed bias.

Once a dispute has arisen, a party may apply to the Court to prevent an arbitrator designated in an agreement from acting on the ground that he is not or may not be impartial even if he knew or should have known this at the time the agreement was made – see S24(1) of the 1950 Act and pages 47-8 *post*.

A shareholder of a party is ineligible – *Sellar* v *Highland Railway* [1919] SC (HL) 19.

A party cannot appoint an employee to arbitrate in a current dispute – Re *Frankenberg and the Security Co.* (1894) 10 TLR 393. Nor is a person eligible if he has exhibited manifest bias against one of the parties.

However, it is not necessarily a disqualification that the arbitrator has worked for one of the parties in some capacity, or that he has performed duties under the contract which may attract criticism. In *Jackson* v *Barry Railway Co.* [1893] 1 Ch. 238, (CA), Barry Railway's engineer was nominated arbitrator in the arbitration agreement and had expressed a view on the very matter in issue in his capacity as engineer. The court held that there being nothing to show he had finally made up his mind this did not prevent the arbitration proceeding. In contrast, in *Nuttall* v *Manchester Corporation* (1892) 8 TLR 513 a City Surveyor who was closely involved with the facts was considered unsuitable and the Corporation was refused a stay.

7. The Limitation Acts

For the purposes of the Limitation Act, 1980, and any other limitation enactment, any term of an arbitration agreement to the effect that no cause of action is to accrue until an award has been made is ignored – S34(2), *ibid*. The limitation period, which is usually six[1] years, therefore runs from the normal date, that is to say, when the right of action or of having the matter arbitrated first arose. A valid notice to arbitrate in accordance with an arbitration agreement stops time running.

S34(3) of the Limitation Act 1980 provides that an arbitration begins when one party serves on the other a notice requiring him to appoint or concur in the appointment of an arbitrator or, where the arbitrator is designated in the arbitration agreement, when one party calls on the other to submit the dispute to him – see per Lord Denning, MR in the *Agios Lazarus* [1976] 2 Ll Rep. 47, at p. 51.

Section 34 applies to statutory arbitrations *mutatis mutandis* – see subsection (6).

The parties may agree to cut down the limitation period, as they do, for instance, with an *Atlantic Shipping* clause, but in certain cases the court may set such a provision aside where it would operate harshly – see pp. 34–6 *post*.

Whether an award has its own vitality will sometimes depend upon the form the award takes, for instance, whether it is declaratory of a right arising under the original contract or substitutes a fresh obligation for an unliquidated claim previously made. In the former case, if the contract is repudiated, terms in it which affect the arbitration may lapse, and it is to be noted that for limitation purposes time will *prima facie* have been running since the original cause of action accrued.

On the other hand, if when an award is made the parties' original rights disappear and those under the award are substituted, a new period of limitation will begin with the award itself – Fletcher-Moulton, LJ, in *Doleman* v *Ossett* [1912] 3 KB 257 at p. 267. By embarking on arbitration proceedings the parties impliedly contract that they will abide by the decision of the arbitrator; on this view time will begin to run again when the implied promise to comply is breached – see *Agromet Motoimport Ltd.* v *Maulden Engineering Co. (Beds) Ltd.* [1985] 2 All ER 436.

An abortive arbitration does not necessarily mean a loss of time for limitation purposes since, when the court sets an award aside or directs an arbitration agreement to cease to have effect in relation to a particular dispute, it may order that time did not run between the commencement of the arbitration and the date of the order – S34(5) *ibid*.

8. Relationship between arbitration and litigation

If one party to an arbitration commences litigation against the other this does not terminate the arbitration nor prevent the arbitrator from continuing to give directions which require compliance. He could make no valid award on matters covered by the litigation without the parties' consent but there is nothing to prevent their giving it. See *Lloyd* v *Wright* [1983] 2 All ER 969, (CA).

Similarly, the grant of a stay does not terminate the action but simply puts it in cold storage; the existence of an arbitration agreement operates as a reason for the court not to exercise its jurisdiction.

An arbitration agreement can only be a defence to an action in the *Scott* v *Avery* type of case, as to which see p. 34, *post*.

It seems from *Northern Regional Health Authority* v *Derek Crouch Construction Co. Ltd.* [1984] 2 All ER 175, that where both types of proceeding are being carried on simultaneously an award dealing with aspects of the dispute which do not form part of the court proceedings can be validly made; an experienced arbitrator can be trusted to refuse to decide issues which overlap. Again, when a plaintiff commences an action and obtains an order for the arrest of a ship despite the existence of an arbitration agreement, the court may stay the action so that the arbitration can proceed without at the same time staying the arrest – see *The Tuyuti* [1984] 2 All ER 545, where the plaintiff was able to show that the defendant might not satisfy any award made against him.

9. The applicable law(s)

Only a brief note on the highly complicated questions which may arise in arbitrations which are not purely domestic is appropriate here.

The law of the main contract governing the substantive rights of the parties, its 'proper law', is a matter for the parties to choose. Where their intention is not clearly expressed in the agreement it will be inferred that they chose the system of law with which the transaction has its closest and most real connection. Among features to be taken into account are the nationality of the parties, the subject matter of the contract, the place of performance, the form of the contract, the language used and the terms of any arbitration agreement, but not the subsequent conduct of the parties. No single feature is sufficiently important to determine the issue though, perhaps particularly in the case of commodity contracts, an agreement that disputes are to be referred to arbitration in a particular country suggests that that country's laws are to govern the contract in the absence of indications to the contrary – a 'sound general rule' but not conclusive, as *Compagnie d'Armement Maritime SA* v *Compagnie Tunisienne de Navigation SA* (1971) AC 572 illustrated. See, also, *Naviera Amazonia* v *Compania Internacional de Peru*, The Independant 11 November 1987.

An arbitration agreement is separate from the main contract even if contained within it and is not necessarily governed by the same law. What that law is also depends on the intention of the parties, whether explicit or to be inferred, and will determine, *inter alia,* the constitution of the tribunal and the validity of, for instance, notices, the reference and the award.

Finally, the actual conduct of the arbitration – the procedure, the rules of evidence, the summoning of witnesses, and the circumstances in which the court may intervene – is governed by the curial law, *lex fori,* that is to say, by the law of the country with which the *proceedings* are most closely connected, almost invariably the place where the arbitration is held.

In *James Miller* v *Whitworth Street Estates Ltd.* [1970] AC 583, a contract for work to be done by Scottish contractors to the Scottish premises of an English company was in standard RIBA form and its proper law was held to be English. However, since a Scottish arbitrator was appointed, Scottish lawyers were instructed and the arbitration took place in Scotland, the curial law was that of Scotland – with the result that the case stated procedure was inapplicable.

For fuller discussions of this topic see the *Miller* and the *D'Armement* cases *op. cit.,* the *Deutsche* case cited on p. 11 and Chapter 4 of Mustill and Boyd.

It is to be noted that the court will be more sympathetic to granting an application for security for costs where English law is the proper law of the contract rather than merely the *lex fori* – see *Bani and Havbulk* v *Korea Shipbuilding and Engineering Corp.* (CA) F.T. Law Reports, 10 July 1987.

It may be added that in England less difficulty arises in regard to foreign law than might otherwise have been expected because it is assumed to be the same as English law unless the contrary is alleged and proved as a fact. Being a matter of fact there is in principle no appeal on such a point.

Finally, it should be noted that the law of the place of enforcement of an award may be of the utmost practical importance.

10. Equity clauses

In *Orion Compania Española de Seguros* v *Belfort Maatschappij Voor Algemene Verzekgringeen* [1962] 2 Ll. Rep. 257, it was said that if the parties required the arbitrator to make his decision by reference to his own ideas of abstract justice there would be no contract since a question of law must remain such for all purposes in view of the courts' 'statutory supervisory jurisdiction over arbitrators'.[2]

The position in regard to evidence has been much more relaxed. An express provision that the tribunal is not bound by the strict rules of evidence is valid and in fact is not uncommon in published procedural rules – see, for instance, County Court Rule 0.19 r. 5(2)(3), and rule 8 of the Refined Sugar Association rules. Again, in *French Government* v *Tsurushima Maru* (1921) 37 TLR 961, the Court of Appeal held that stipulating for an arbitration to be by 'commercial men' indicated that neither normal court procedure nor the strict rules of evidence might be followed.

In *Deutsche Schachtbau – und Teifbohrgesellschaft mhb* v *Ras Al Khaimah National Oil Co.* [1987] 2 All ER 769, CA, the (Swiss) arbitrators were given the right to apply whichever law they deemed appropriate and were held entitled to opt for 'internationally accepted principles of law governing contractual relations'. The criteria to be satisfied in such a case are that the parties intended to create legally enforceable rights and obligations, that there is sufficient certainty, and that there is no breach of public policy. The award was held enforceable forthwith under S26 of the 1950 Act as a Convention Award – see Chapter 9 and S36(1) of the 1950 Act.

Perhaps the case where the wording is closest to that of entitling the arbitrator to act as 'amiable compositeur' is *Eagle Star Insurance Co. Ltd.* v *Yuval Insurance Co. Ltd.* [1978] 1 Ll. R. 357, where the Court of Appeal held an arbitration agreement in the following terms to be valid:

'The Arbitrators and Umpire shall not be bound by strict rules of law but shall settle any difference referred to them according to an equitable rather than a strictly legal interpretation of the provisions of this Agreement'.

In as far as this decision can be interpreted as merely giving the arbitrator authority to ignore 'strict interpretations', a reconciliation between the *Eagle Star* and *Orion* cases is possible. It may, however, be that the ability of the parties to extend the powers of an arbitral tribunal now goes considerably further and enables them to permit, indeed to direct, an arbitrator to act as 'amiable compositeur'[3], this notwithstanding Mustill and Boyd's doubts as to how logically to interpret the substitution of ideas of fairness for strict legal rights – see pp. 605–617 of that work.

If logic is to be brought into it, it is by no means clear, however, why parties may not validly agree to have their rights and obligations determined by the ideas of fairness of a specified or ascertainable individual since it is well-established that an arbitrator may make, or alter, the terms of the contract itself. Nor is the latter power necessarily confined to the future; the Court of Appeal decision in the *Northern Regional Health* case – see page 9 *ante* – that an arbitrator may be enabled to substitute his own views for those in an architect's certificate had the effect of altering what would otherwise have been the parties' position in law. It is indeed submitted, though with very considerable diffidence in view of *Orion,* and Mustill and Boyd's reservations, that in spite of possible difficulty in defining the precise limits of the arbitrator's powers an equity clause would now be treated as valid and take effect as a contract to abide by the arbitrator's award. So far as appeals are concerned (in the absence of an exclusion agreement) it is suggested that the court might treat the arbitrator's views as to what was fair as final unless of the opinion that the true and only reasonable conclusion of what was fair contradicted his determination, in other words, by analogy with *Edwards* v *Bairstow* principles – see [1956] AC 14 – and naturally subject also to arguments based on technical misconduct, bias, breaches of natural justice, and so on.

Some small support for the view that an arbitrator may determine the rights of the parties on the basis of his views of abstract justice where the parties purport to give him the necessary power is that the ratio of the *Orion* case by reference to the supposed general supervision of the courts over arbitrators has now been negatived. Besides, other considerations which may have influenced the court in that case have gone with the abolition of the Case Stated system and of jurisdiction to set an award aside on the ground of errors of fact or law on its face – not to speak of the creation of power to enter into exclusion agreements.

11. Statutory arbitrations

Since this book is concerned with arbitrations of a consensual character statutory arbitrations are strictly speaking outside its scope, but a few words on the subject may be helpful.

(a) Reference in specific cases

A considerable number of statutes provide for disputes of particular kinds to be referred to 'arbitration'. One illustration is S76 of the Friendly Societies Act, 1974, which provides *inter alia*, that disputes between (i) a member and an official, (ii) a society and a branch, and (iii) branches, are normally to be determined in accordance with the society's or branch's rules rather than the courts. Again, there are various provisions in the Housing Act 1957, in relation to compensation and other causes of disagreement.

(b) Special rules

Some of the statutes contain their own procedural codes in greater or less detail and/or provide that the Arbitration Acts are to apply to a given extent.

(c) Residual application of the Arbitration Acts

With the exceptions mentioned below and subject to any modifications made by other statutes the provisions of Part I of the 1950 Act and of Ss1–6 of the 1979 Act apply to arbitrations under other Acts as though they were arbitrations pursuant to an arbitration agreement – see S31 of the 1950 Act, as amended by S8(2)(d) of the 1975 Act, and S7(1)(d) of the 1979 Act. The exceptions are the following provisions of the 1950 Act—Ss2(1) and 3 (the effect of a party's death and bankruptcy, respectively); S5 (interpleader issues); S18(3) (agreements to share arbitration costs avoided if made before a dispute arises); Ss24 and 25 (power of court where arbitrator is not impartial or fraud is in issue and where an arbitrator is removed or his authority revoked); S27 (power to over-ride *Scott* v *Avery* clauses); S29 (re deposits with wharfingers).

Exclusion agreements have no application to statutory arbitrations – S3(5) of the 1979 Act.

(d) General comments

Section 34(6) of the Limitation Act 1980 provides that that section also applies to statutory arbitrations. See para. 7 *ante*.

The proceedings of a statutory arbitration, unlike those of a consensual arbitration, are subject to the courts' inherent right of supervision which it possesses in the case of inferior courts and tribunals.

A rather wider treatment of statutory arbitrations will be found in Russell at pp. 10–21.

Notes

1. Twelve, where the relevant document is under seal.
2. The courts may intervene in the case of manifest misconduct, but the existence of a general supervisory jurisdiction has since been denied – see para 8-6 post.
3. Article 28(3) of the Unitral Model Law contemplates that the parties may do just that, as does the new Dutch arbitration law.

2 Advantages and disadvantages of arbitration as a means of settling disputes

While an arbitrator's personality and skills will inevitably tend to be more important to the disputants than those of a judge, regardless of the composition of the tribunal, there are circumstances which suit one procedure rather than the other. Taking a number of different aspects one may say that, generally speaking and with exceptions in individual cases, arbitration tends to be preferable to litigation from the point of view of ten of them – see pp. 14–20 below – and litigation generally better in the case of three – see pp. 20–22 – while in two the advantages depend on the circumstances – see p. 23.

Arbitration generally preferable

1. Cost

The cost of resolving a dispute depends largely upon its nature – the complications, any special skills needed, the amount at stake and, above all, how long it takes.

Arbitration has one built-in cost disadvantage, namely, that the arbitrator himself has to be paid.[1] In addition a suitable venue must be found but this does not necessarily cost more than the Court fees payable in litigation.

In other respects, arbitration is usually less expensive, in part because substantial economies of time are possible, for example, in the reading of documents and the giving of background explanations. Also, where Counsel are briefed, their clerks may possibly accept a little less than they would expect for a court hearing, in which event there may be some corresponding diminution in solicitors' costs.

More important, the arbitrator's earlier and closer involvement with the case often enables him to reduce the area of dispute, if not resolve it altogether, in cases where otherwise only the fear of crippling costs would induce a settlement.

The main expense in litigation is that of the lawyers employed. Unfortunately, the charges of solicitors' litigation departments are often high in relation to the skill and ingenuity supplied and since the general public and many of its advisers are only dimly aware of arbitration as a practical alternative, innumerable sound claims are abandoned each year because of the threatened costs or the past shortcomings of other members of the profession.

A rather more fundamental question is whether, or to what extent, it is practicable to limit legal representation or even dispense with it altogether.[2]

There can be no doubt that the customary ability of lawyers to present the facts and their arguments logically, and to identify and weed out issues of lesser importance, can save much time, particularly in complicated cases. On the other hand, many professional arbitrators, particularly in the construction and maritime fields, consider that introducing lawyers into technical disputes merely causes delay and an escalation of costs without countervailing advantages.

Perhaps the most one can say is that the necessary extent of lawyers' participation may be rather less than is generally assumed. However, their involvement is unlikely to lessen more than marginally in the foreseeable future. In practice, parties are apt to fly to their lawyers as soon as a dispute arises and it is not every solicitor who recognizes – or would acknowledge if he did – that occasionally the clients' interests would be best served if he and his opposite number nominated a suitable arbitrator and quietly withdrew.

Although comparative cheapness *is* traditionally regarded as one of the main benefits of arbitration exceptions are so numerous that it might have been better to relegate 'cost' to the third category (where advantages or the reverse depend on the circumstances) except for one final consideration, namely, that appeals from an arbitrator are much less common, especially since the 1979 Act came into force and the guidelines in regard to granting leave to appeal were laid down in *The Nema* – see Chapter 8 para. 3(g).

2. Lack of Publicity

Arbitration proceedings are held in private and even the existence of a dispute, let alone its nature, seldom becomes generally known.

As indicated above, appeals are comparatively rare but, even so, much of the evidence which could become public knowledge with a full court hearing never comes to light. In any case, no publicity is likely to be attracted unless the appeal involves a point of sufficient interest to warrant its being reported.

Unfortunately, until practice under the 1979 Act finally settles down, the mere granting, or refusal, of leave to appeal may be such a point. So may challenges to the reference, for instance, on the ground that there is no defence to summary proceedings brought under Order 14, or that the nature of the case is such that arbitration is unsuitable. The latter contention is much less common than hitherto because arbitration is more widely accepted as a satisfactory procedure and as one which once chosen by the parties should normally be pursued.

An action to remove an arbitrator on the ground of bias or misconduct does not of course enjoy the privacy of the arbitration proceedings themselves.

Quite apart from minimising the risk of disclosure to third parties and adverse publicity generally, the resolution of a dispute in private can be much less disruptive of the parties' business relationship.

3. Speed

Arbitrations normally take less time than court proceedings though international cases with three arbitrators can be very lengthy.

Even where the speedy Order 14 procedure is available because there is no defence, an arbitrator is usually *capable* of acting still more quickly especially if the court proceedings have to be served abroad. An exhaustive analysis of the pro's and con's of the two procedures will be found in Mustill and Boyd at pp. 442-3. See also para. 12, *post*, at p. 22.

In *Northern Regional Health Authority* v *Derek Crouch Construction Co. Ltd.* [1984] 2 All ER 175 (CA) the Master of the Rolls considered that the Official Referee was right to take account of the fact that the arbitrator could hear a matter forthwith while a court hearing would have taken another year.

Only when it comes to the decision is litigation normally quicker. A judge frequently gives an extempore judgment when the case terminates; regrettably, and sometimes reprehensibly, some arbitrators take weeks, even months, to make their awards.

An important cause of delay in litigation is the haphazard way in which lawyers and the courts tend to work. Judges are usually allotted cases at a late stage and progress during the interlocutory period depends mainly on the solicitors who are apt to be very busy and to put on one side cases which are difficult or where the client exerts less pressure. Also, being frequently in similar trouble themselves, they tend to be over-indulgent when the other side asks for further time. NB The comments made above on Court procedure do not apply to anything like the same extent to the Commercial or Official Referee's Courts – see, in particular, Practice Note of 23/10/87 [1987] 3 All ER 799 which, *inter alia*, requires written submissions to be submitted in advance.

The arbitrator on the other hand will usually have been familiar with the issues in each of his cases since its earliest interlocutory phase. It represents an open file the early closing of which is a desirable end in itself; if a pleading is late, as like as not he will try to find out why and hurry it along, though it must be conceded that some arbitrators are content to adopt a much more passive role.

In the case of a hearing, pressure from the arbitrator to fix a provisional date at an early stage, and keep to it so far as reasonably possible, can cut out much delay.

In this connection, it must be remembered that it may be to one party's advantage to delay resolution of a dispute as long as possible. The other party should be on the look out to frustrate any tactics directed to this end and the arbitrator should do his best to help him.

A neat alternative to the rather clumsy weapon of applying to the court to remove an arbitrator or umpire who is not reasonably expeditious – see S13(3) of the 1950 Act – is available where the LMAA Terms (1987) apply; a party can force a sole arbitrator to retire if he cannot offer a date for the hearing within a 'reasonable' time – defined as four months in the case of a hearing estimated to last not more than two days – and, if the parties cannot agree on a substitute, the President of the Association will appoint one.

4. Technical expertise

It has long been recognised that normal court procedures are not suited to very complicated disputes and such cases are usually transferred to the Official Referee. Such proceedings benefit from the institution of a user's committee and are notably efficient, but as they are not held in private and tend to be more costly it may be preferable to employ a technically equipped arbitrator. Technical knowledge is not of itself sufficient of course. An expert with a bee in his bonnet and lacking a judicial approach is worse than any Judge, however ill-instructed in technology. As the present Master of the Rolls said recently: 'An unskilled arbitrator is an unmitigated disaster'.

Fortunately, in most fields there is a supply of technical experts with an adequate knowledge of arbitration. The relevant professional body is likely to have a list of members with arbitral experience who are willing to be nominated. Approaching the matter from the other end, the Chartered Institute of Arbitrators maintains panels of members who are qualified in one or more disciplines in addition to that of arbitration.

The possession of technical expertise may result in a trap if the arbitrator is unwary, namely, that of reaching conclusions on the basis of his own experience without letting the parties know the way his mind is working, thereby depriving them of the chance to argue in favour of other views. Where this happened in *Zermalt Holdings SA* v *Nu-Life Upholstery Repairs Ltd.* [1985] 2 EGLR 14 (CA), it was held contrary to the rules of natural justice; the arbitrator was removed and his award set aside.

5. The enforceability of an award

A valid award can be made enforceable as a judgment of the court under S26 of the 1950 Act which may be a little less clumsy than having to bring an action to enforce a decision of an expert which it has been agreed shall be binding. A pronouncement by a conciliator is not enforceable at all.

An arbitrator's award may in fact have an important advantage over a judgment in that its enforceability aboard, and that of a foreign award in the UK, depend upon whether the foreign country has acceded to the New York Convention on the Recognition and Enforcement of Foreign Arbitral Awards, 1958. The enforceability of a foreign judgment on the other hand depends upon the existence of a specific provision in the law of the country where enforcement is sought. The latter is a narrower network and may be less easy to establish in a given case. This subject is dealt with in greater detail in Chapter 9.

6. Representation

In the High Court a litigant may appear either in person or by Counsel. The latter must be instructed by a solicitor who will normally attend the hearing, though he is likely to be represented by a clerk in an unimportant case. However, since a company cannot appear in person it is obliged to instruct Counsel, a needless addition to the high cost of litigation which is particularly regrettable in the case of the 'one-man' company.

In an arbitration no such constraints exist and a party can appear by himself or by any professional or indeed non-professional person, including in the case of a corporation, a director, secretary or any other officer or adviser.

7. Extent of jurisdiction

The English Courts will not make a bargain for the parties – see *Scammell* v *Ouston* [1941] AC 251 – which can be a considerable disadvantage in the world of commerce. Fortunately, however, it is now firmly established that an arbitrator may wield the power the courts lack provided it is given him either expressly or by necessary implication – see, for instance, per Lord Denning, M.R., in *Courtney* v *Tolaini* (1975) 1 All ER 716, at p. 719g. This power can, for instance, extend to fixing the price for goods, altering the terms of a contract to meet changed circumstances or, in relation to real property, setting out a right of way or deciding the terms of a lease.

In *F. & G. Sykes (Wessex) Ltd.* v *Fine Fare Ltd.* [1967] 1 Ll. Rep. 53 (CA) a five-year contract for the sale of poultry specified how many should be supplied in the first year; for the remaining four it was to be as many as should be agreed, but no agreement was reached. The contract provided that any difference as to the meaning or effect of the main contract, or as regards the performance by either party of its obligations under it, or in relation to any matters incidental thereto should be submitted to arbitration; it was held that the arbitrator had jurisdiction under these provisions to decide how many chickens should be supplied.

In *Northern Regional Health Authority,* v *Derek Crouch Construction Co. Ltd.* [1984] 2 All ER 175 (CA) the arbitration agreement gave the arbitrator power to vary an architect's certificate and substitute his own opinion. It was held that in litigation[3] as distinct from arbitration the court would have had no power to vary the certificate.

8. Convenience

While the courts endeavour to meet the convenience of the parties so far as they reasonably can, the emphasis in arbitrations is rather different. There the convenience of the parties is normally paramount in considering administrative questions of whatever kind. This can extend to the taking of evidence as well as the timing and situs of the hearing.

9. Flexibility of procedure

There can be no doubt that most arbitrators are decidedly more flexible in procedural matters than the courts and this is generally speaking a great advantage.

There are a number of possible reasons – the fact that the arbitrator is master of the proceedings before him and that there is more than one set of acceptable procedural rules to cull from provided both parties agree; that he will be much less well grounded in the RSC and County Court Rules than a judge; that he may happen to be impatient of legal formalities and/or may have some familiarity with procedures in Civil Law countries, and so on. It might even be argued that a person who lacks creativity or is emotionally or intellectually unduly wedded to court procedure should exercise his gifts outside arbitration.

Examples of procedural devices which arbitrators adopt to advantage more readily than courts are 'documents only' proceedings and the determination of liability before quantum. In addition, the inquisitorial approach employed in civil law countries can be used to a limited extent in some circumstances, or in regard to certain specific issues, where it seems advantageous – see Chapter 6.

Naturally, arbitrators occasionally go off the rails, but recourse to the courts usually remains possible where the lapse is a serious one.

10. Relationship between tribunal and disputants

Where the parties can agree on the choice of an arbitrator whom they know their confidence in him and his knowledge of them may be mutually beneficial. There may be comparable advantages where the nomination is by the President of some professional or trade body which is connected with the field in which the differences have arisen. A judge whose background does not give him a 'feel' for a situation may find it much more difficult to reach a solution which satisfies either him or the parties.

Litigation generally preferable

11. Multi-party disputes

Disputes involving more than two parties tend to end up before the courts for technical reasons. For one thing, arbitration proceedings normally arise pursuant to a clause in a contract between two parties and neither they nor the arbitrator have any power to force a third party (or parties) to join in. Similarly, however close a third party's involvement with the facts, unless everyone concerned agrees he has no means of participating or otherwise intervening – even if he actually knows the arbitration is taking place, which may well not be the case.

In *Oxford Shipping Co. Ltd.* v *Nippon Yusen Kaisha, The Eastern Saga* [1984] 2 Ll. Rep. 373, Leggatt, J., held, disapproving remarks to the contrary in Mustill and Boyd (at p. 112), that in the absence of the consent of all concerned even the court cannot order that two arbitrations are to be held concurrently or in consonance with each other let alone that they should be consolidated however convenient such a course might seem to be.

The possible complications are well illustrated by *Interbulk Ltd.* v *Aiden Shipping Co. Ltd. (The Vimeira)* [1984] 2 Ll. Rep. 66 which, although concerned with only three parties – an owner, and time and voyage charterers – and relatively simple facts, went to some 10 court hearings in addition to various hearings before two arbitral tribunals. Even having the same arbitrator for two consecutive arbitrations, assuming that this can be arranged, does not necessarily solve the problem – see *Abu Dhabi Gas Liquefaction Co.* v *Eastern Bechtel Corp.* [1982] 2 Ll. Rep. 425, (CA); the evidence in the two arbitrations may not be the same.

The throwing away of costs is not the only problem which lack of a power to consolidate may engender; the possibility of two tribunals reaching different conclusions leaving an unfortunate party in the middle is likely to be much worse.

Certain sets of procedural rules contemplate a very limited form of consolidation. For instance, in relation to multi-party disputes Section (C) of the First Schedule to the LMAA Terms (1987) gives the arbitrator power (i) to hear concurrently all disputes which have been referred to him, and (ii), on the application of a party, to join any third party or parties who have given written consent to be joined. Again, the Refined Sugar Association Rules deal with chains of claims by providing that where the subject-matter and terms of contracts are identical except as to date and price, and all parties agree in writing, the arbitration may be between first seller and last buyer with all intermediate parties being bound by the award – see p. 68 *post*.

Just occasionally, contracts provide for the joinder of other parties in certain circumstances, but this could not be enforced against an unwilling third party and would not always be desirable. Amongst other things, it would reduce the degree of privacy – one of the points taken in the *Vimeira* case *op. cit* – and might prejudice one or more of the parties by causing delay or additional cost.

Even if two arbitrations arising out of the same facts are held concurrently, separate awards must be given unless the parties have agreed to amalgamate them – see *Wilhelmsen* v *Canadian Transport Co.* (*The Takamine*) [1980] 2 Ll. Rep. 204, at p. 208.

See also paras 6(2)–8 of the illustrative Award on p. 117.

There has been some reluctance to press for a change in the law permitting the court to order consolidation – as recent Hong Kong legislation has done – amongst other reasons because it would involve overriding the wishes of the parties and could cause difficulty in enforcing the award internationally.

In litigation, by way of contrast, there is usually no problem in consolidating two or more actions, or in joining other persons by means of third party proceedings.

12. Strength of the case

Where a claimant considers he has a very strong case, that is to say, one to which the other side has no real defence, he will no doubt prefer the summary court procedure provided by Order 14 of the Rules of the Supreme Court. Unlike ordinary actions this procedure, where applicable, is comparatively both cheap and swift. The court also has power under Order 29 to order a defendant to make an interim payment, for example of an amount admitted to be due.

An arbitrator would be unable to follow a similar procedure unless given specific power to do so and even then there might be difficulty in enforcing the resulting award abroad if that became necessary.

However, an arbitrator can achieve some approximation to Order 29 in certain cases by making an interim award – see para 7–3 *post* at p. 74.

13. Reference to the European Court of Justice

Since an arbitrator cannot himself refer questions of EEC law to the Court of Justice litigation may be preferable to arbitration if the need for such a reference is anticipated. If the matter were arbitrated it would be necessary either to proceed under S2(1) of the 1979 Act or to appeal under S1(3) – see *Bulk Oil (Zug) AG v Sun International Ltd.* [1984] 1 All ER 386.

Which is better depends on the circumstances

14. Nature of the dispute – whether of fact or law

Disputes of fact, especially those of a technical or complicated character, tend to be more suited to arbitration.

It is to be noted that a number of professional and trade associations have recognized that the ordinary courts system is ineffective in coping with small claims in the field that concerns them owing to the excessive legal costs, or the lack of specialist Registrars, or both, and have set up efficient arbitration arrangements of their own. The latest to do so is the Law Society with assistance from the Chartered Institute of Arbitrators. See, also 'Documents only' arbitrations at p. 66.

On the other hand, a decision of the court is usually preferable where the issue is one of law, for example, the true meaning of a phrase in a document, especially where the party concerned contemplates using a similar document again and desires a decision which can be cited to third parties. In this connection, it is as well to remember that the court normally confines comments to the actual facts before it and is most reluctant to make general declarations.

There is, however, one practical point to note. If the main issue is one of law a party with little confidence in his arguments may opt for arbitration, if open to him to do so, in the hope of a bad decision in his favour since, in practice, an appeal by the other side will probably be impossible for reasons set out in Chapter 8.[4]

15. Legal Aid

Legal aid is not available in arbitration cases, a circumstance which may be either an advantage or disadvantage according to the point of view.

Notes

1. In one recent case, the relevant Rules required five arbitrators each of whom charged £75 per hour and there were heavy legal costs on top because the arbitrators required a lawyer to be present throughout. The court reduced the latter charges but otherwise upheld them.

2. A company's costly disadvantage in litigation that it must brief counsel in the High Court is noted in the section on 'Representation' at p. 18.
3. This is a different case from *Neale* v *Richardson* [1938] 1 All ER 753, (CA), where an architect whose certificate was a condition precedent to payment refused to act as arbitrator and it was held that the court had jurisdiction to determine how much was due.
4. There is a rather similar rule of thumb in tax appeals – appeal to the General Commissioners with a bad case and the Specials with a good one.

3 The arbitration agreement

1. Its form and contents

S32 of the 1950 Act defines an arbitration agreement as 'a written agreement to submit present or future differences to arbitration, whether an arbitrator is named therein or not'[1] It would therefore cover a clause which is ancillary to the main subject matter of a contract, a separate agreement relating to arbitration, and an ad hoc submission of a dispute which has already arisen.

The naming of the arbitrator (or not) and his appointment are dealt with in the next Chapter. Here are discussed the form of an arbitration agreement, the legal effect of different wordings and the requirement that it must be in writing.

So far as form is concerned it can range from a very simple phrase such as: 'Any dispute arising out of this agreement shall be referred to arbitration', to wordy extravaganzas. Best is a wording somewhere in between but taking into account the considerations mentioned below.

First, the intention should be clear although a high degree of particularity is not obligatory – see, for instance, *Hobbs Padgett & Co. (Reinsurance) Ltd. v J. C. Kirkland Ltd.* [1969] 2 Ll. Rep. 547 (CA), where a provision that there should be a 'suitable arbitration clause' was held to constitute a valid arbitration agreement. 'Arbitration to be settled in London' has also been held sufficient – *ibid* at p. 549.

Second, the arbitration agreement should state either who is to arbitrate, with alternatives if he cannot or will not act, or how he is to be chosen.

Third, it is generally speaking a considerable advantage to refer to a set of rules as being those in accordance with which the arbitration is to be conducted. See Chapter 5 p. 51–2.

Finally, it should be wide enough to cover each type of dispute the parties would wish to refer to arbitration should it arise.

If, as is usual, it is desired to include torts connected with a contract, 'arising out of', which is probably wider than 'arising under', would be sufficient for the purpose; it would also cover a dispute as to the construction of the contract – *Thorburn* v *Barnes* [1867] LR 2 CP 384. 'Arising out of or in connection with' is wider still. In *Astro Vencedor Compania Naviera SA* v *Mabanaft GmbH, The Damianos* [1971] 2 QB 588 (CA) 'arising during the execution of this contract' was held to include claims for tort if sufficiently close to the performance of the contract, in that case the wrongful arrest of a ship.

However, where a contractor brought an action for damages alleging that he had been induced to enter into a contract by fraudulent misrepresentation, the defendant failed to obtain a stay though all disputes 'upon or in relation to or in connection with' the contract were to be arbitrated – *Monro* v *Bognor UDC* [1915] 3 KB 167, (CA).

An agreement to refer disputes as to 'the meaning and effect' of a charterparty was held to cover applying it to facts which had arisen, such as the amount of hire payable, but not to give the arbitrator jurisdiction to award damages – *Richards* v *Payne & Co.* (1916) 86 LJKB 937. A claim for rectification of a contract does not 'arise in relation to these presents' – *Printing Machinery Co. Ltd.* v *Linotype and Machinery Ltd.* [1912] 1 Ch. 566.

An arbitration agreement forming part of a wider agreement, where the parties are *ad idem* as to who should arbitrate, might consequently be on the following lines: 'Any dispute arising out of or in connection with this contract shall be referred to Mr X or, in the event of his being unable or unwilling to act, to an arbitrator to be agreed between the parties or, failing agreement within 14 days of one party requesting the other in writing to concur in a specified appointment, to be nominated by the President for the time being of the Institute of Chartered Accountants in England and Wales or his duly authorised deputy. Such arbitration shall, unless the arbitrator with the consent of the parties decides otherwise, be conducted in accordance with the most recent[2] edition of the Rules of the Chartered Institute of Arbitrators'.

For additional provisions which it is wise to add where one of the parties is foreign, or non-resident, see Chapter 9, pp. 104-6.

There is no statutory provision which requires an arbitration agreement to be signed. Nor need the agreement be contained in a single document; it is sufficient that a document or documents exist which recognize or confirm that there is an agreement to submit – *Excomm Ltd* v *Ahmed Abdul-Qawi Bamaodah* [1985] 1 Ll. Rep. 403 (CA). Provided an arbitration agreement forms part of a written agreement between the parties, the oral assent to the contract by one of the parties is sufficient – *Zambia Steel and Buildings Supplies Ltd.* v *James Clark and Eaton Ltd.* FT Comm. LR 15 August 1986. NB Lack of the signature 'of the person to be charged or some person lawfully authorised thereunto' may prevent enforcement of the award where S4 of the Statute of Frauds 1677 or S40 of the Law of Property Act, 1925, applies.

Where a party may elect by notice within a given period to refer to arbitration a dispute which would otherwise go to the courts, an arbitration agreement comes into existence upon such election being duly exercised in writing – *Westfal-Larsen* v *Ikerigi Compania Naviera SA* [1983] 1 All ER 382.

The earlier view that lack of mutuality renders an arbitration agreement invalid is no longer good law – see *Pittalis* v *Sherefettin* [1986] 2 All ER 227 (CA).

An *oral* agreement to arbitrate is valid as a common law arbitration but suffers from grave defects: very little of the legislation is applicable since nearly all the references are to 'arbitration agreements' as defined – see p. 25 – and either party may withdraw at any time.

2. Implied terms

The 1950 Act deems an arbitration agreement to contain the following provisions unless it expresses a contrary intention:

(a) That there is to be a single arbitrator – S6;

(b) That where the reference is to two arbitrators they may appoint an umpire at any time and shall do so forthwith if they disagree – S8(1);

(c) That if there are two arbitrators and they notify any party or the umpire, in writing, that they disagree, the umpire may enter on the reference forthwith in place of the arbitrators – S8(2);

(d) That the parties and all persons claiming through them shall, subject to any legal objection, (i) submit to being examined on oath or affirmation, (ii) produce all documents within their possession or power which may be called for, and (iii) do all other things during the arbitration proceedings which the arbitrator or umpire may require – S12(1);

(e) That witnesses are to be examined on oath or affirmation if the arbitrator or umpire thinks fit – S12(2);

(f) That an interim award may be made – S14;

(g) That the arbitrator or umpire is to have the same power as the High Court to order specific performance of any contract not being one relating to land or any interest in land – S15;

(h) That the award is to be final and binding on the parties and on parties claiming under them – S16;
(i) That a clerical error or a mistake arising from a slip in an award may be corrected – S17;
(j) That the costs are in the discretion of the arbitrator or umpire, and that he may tax them, with power to award costs as between solicitor and client – S18(1). See now 0.62 of the RSC and Chapter 7 para. 8(b) pp. 80-1; and
(k) That certain simple interest may be awarded – S19(A). See p. 85.

Terms are also implied by the general law of contract, for example, that the parties will co-operate in the expeditious conduct of the reference – *Paal Wilson & Co. A/S* v *Partenreederei Hannah Blumenthal* [1983] 1 AC 854 – that they will honour any award made – *Agromet Motoimport Ltd.* v *Maulden Engineering Co. (Beds) Ltd.* [1985] 2 All ER 436 – and that strangers are to be excluded from the hearing and conduct of an arbitration under the agreement – *Oxford Shipping Co. Ltd.* v *Nippon Yusen Kaisha* [1984] 3 All ER 835. This last mentioned implied term is one reason for barring the consolidation of cases, in the absence of consent of the parties.

3. Enforcing an arbitration agreement

Lord Denning, MR, said in *Eagle Star Insurance Co. Ltd.* v *Yuval Insurance Co. Ltd.* [1978] 1 Ll. R 357 at pp. 361-1, that the courts welcome arbitrations in commercial disputes and encourage references: 'It seems to me that if a defendant who is being sued in the Courts asks that a matter should go to arbitration in accordance with their agreement, *prima facie* that agreement ought to be honoured: the action should be stayed . . .'

The court will not however grant specific performance of a contract to arbitrate, and it only constitutes a defence to an action in the *Scott* v *Avery* and *Atlantic Shipping* types of case, that is to say, where the plaintiff has adopted a procedure, or over-stepped a time limit, contrary to what the parties had agreed – see pp. 34-6.

It seems that a foreign judgement obtained in breach of an arbitration agreement will not be enforced in England with the result that the efficacy of such an agreement is indirectly enhanced.

If one party to an arbitration agreement brings an action, the other may enforce the agreement by applying for a stay – see p. 30. Alternatively, an aggrieved party may take advantage of the procedures provided by the 1950 Act. Thus, depending on the circumstances, he may himself appoint an arbitrator under S7 or apply to the court under S10 – see Chapter 4 paras 1 and 4 respectively. Once appointed, an arbitrator has various ways of combating lack of cooperation including power to proceed *ex parte* see p. 58.

Except where an *Atlantic Shipping* clause imposes a limitation – see *post*, at pp. 34–6, failure to pursue a claim timeously does not extinguish the enforceability of an arbitration agreement, since, unlike the court, an arbitrator cannot dismiss a claim for want of prosecution unless power to do so has been conferred upon him, for example, by procedural rules which govern the arbitration, see, e.g. Rule 2:8 of the GAFTA Rules which provides, *inter alia*, for a claim to lapse if not pursued within one year. The court has inherent jurisdiction to dismiss a claim before it for want of prosecution to guard against misuse of its process but an arbitrator has no need of such power since in arbitration both parties owe each other a duty to pursue it with expedition and the complaining party could always have called on the arbitrator to issue directions. Nor can the court itself exercise such a power in an arbitration – *Bremer Vulkan Schiffbau and Maschinenfabrik* v *South India Shipping Corporation* [1981] 1 Ll. Rep. 253 (HL). Followed in *Food Corporation of India* v *Antclizo Shipping Corp.* [1987] 2 Ll. Rep. 130 (CA).

4. Resisting an arbitration agreement

On the other hand, in certain circumstances, one party can obtain an injunction preventing the other from proceeding with an arbitration, for example:

(a) Where the arbitration agreement does not fully cover the dispute. In *The Phonizien* [1966] 1 Ll. Rep. 150, a party obtained a declaration that an arbitration clause in a charterparty did not apply to a bill of lading arising under the charter and that consequently an arbitrator appointed to deal with a dispute arising under the bill had no jurisdiction to do so;

(b) Where the contract containing the arbitration agreement is *prima facie* invalid;

(c) Where one party has done something which indicates an offer to abandon the reference, and there is evidence of an acceptance of that offer. Mere failure to act, however, does not indicate an intention to relinquish a right to refer to arbitration. Nor does it result in equitable estoppel since mere silence and inactivity are equivocal – *Allied Marine Transport Ltd* v *Vale do Rio Doce Navegaçao SA, The Leonidas D.* [1985] 2 All ER 796, (CA) where it was held that there was no evidence that in failing to press their case the owners were acting on any representations made by the other side;

(d) Where the facts show that the arbitration agreement was repudiated by mutual consent, as in *Andre et Cie SA* v *Marine Transocean Ltd. ('The Splendid Sun')* [1981] 2 Ll. Rep. 29, (CA);

(e) Where the agreement to arbitrate has been frustrated by some external cause, but not by mere delay since each party is under an obligation to the other to take steps to call on the arbitrator to issue directions; failure in this regard therefore cannot result in 'frustration' – *Paal Wilson & Co. A/S v Partenreederei Hannah Blumenthal* [1983] 1 AC 854; or

(f) Where the claim is statute-barred.

An alternative way of blocking an arbitration which may sometimes be appropriate is to sue for a declaration – for example, to the effect that what has been referred is not within the arbitrator's jurisdiction.

5. Applications for a stay

(a) Statutory provisions

An indicated above, if a party (or someone claiming through or under him) breaches an arbitration agreement by bringing an action in the courts the other side's remedy is to apply for the proceedings to be 'stayed', that is to say, halted – see S4(1) of the 1950 Act and S1(1) of the 1975 Act.

These sections may be compared and contrasted:

(i) S1(1) of the 1975 Act has no application to a 'domestic arbitration agreement', that is to say, one which does not contemplate arbitration taking place abroad, *and* no party to which at the time the proceedings are commenced[3] is *either* a national of or habitually resident in a State other than the UK, *or* a company incorporated in, or whose central management and control is in, such a State – see S1(4) *ibid*.

S4(1) of the 1950 Act is of general application except that it is excluded where the other section applies – S1(2) of the 1975 Act.

(ii) The court's power under the earlier section is discretionary while the terms of the later section are mandatory. However, in both cases (i) the proceedings must relate to a matter agreed to be referred, and (ii) the application to stay must be made after appearance but before the applicant has taken any step in the proceedings – see p. 32.

(iii) S4(1) of the 1950 Act states that before granting a stay the court must be satisfied positively that there is no sufficient reason why the matter should not be referred – see p. 32 – and that at the time the proceedings were commenced the applicant was, and still remains, ready and willing to do whatever is necessary for the proper conduct of the arbitration.

S1(1) of the 1975 Act, by way of contrast, provides that a stay *is* to be granted unless the court is satisfied as to quite different matters, namely, that the arbitration agreement is null and void, inoperative, or incapable of performance, alternatively, that there is no dispute between the parties.

(iv) Although, on the face of it S1(1) of the 1975 Act seems to apply only where a party to an arbitration agreement which is already in existence brings proceedings, the court in *The Tuyuti* [1984] 2 All ER 545 (CA), seeing no reason for such a limitation, prayed in aid the terms of the New York Convention to resolve the 'ambiguity', and granted a stay on the basis of an ad hoc arbitration agreement entered into after the action had begun.

(v) The High Court may refuse a stay where it has power under S24 of the 1950 Act either to give leave to revoke the authority of an arbitrator or umpire nominated or delegated in the agreement *or* to order that the arbitration agreement is to cease to have effect. These matters are dealt with at pp. 47-8 *post*.

(b) Other considerations

(i) A party cannot apply for a stay and at the same time challenge the existence or validity of the contract which contains the arbitration agreement. Thus in *Toller v Law Accident Insurance Society Ltd.* [1936] 2 All ER 952 (CA) the respondent denied that a policy had been issued but sought, in vain, to rely on the arbitration clause which it would have contained if it had been issued.

(ii) It may be appropriate for only part of the proceedings to be stayed, namely, where they are divisible and only part of the dispute falls within the arbitration agreement. Similarly, where a claim in Order 14 proceedings is admitted in part the claimant is entitled to judgment to that extent but a stay will be granted on other aspects of the claim if there appears to be an arguable defence - see *Ellis Mechanical Services Ltd. v Wates Construction Ltd.* [1978] 1 Ll. Rep. 33, (CA). In *Nova (Jersey) Knit Ltd v Kammgarn Spinnerei GmbH* (1977) 2 All ER 463, the House of Lords came to the same conclusion in regard to a non-domestic agreement. However, where the claim is for an *un*liquidated sum any stay must be of the whole action - *Associated Bulk Carriers Ltd. v Koch Shipping Inc. The Fuohsan Maru* [1978] 2 All ER 254 (CA).

(iii) A stay will not be granted where there is no arbitration clause in a parallel agreement for which the applicant was responsible - *Bulk Oil (Zug) AG v Trans-Asiatic Oil Ltd.* [1973] 1 Ll. Rep. 129.

(iv) S12(6)(f) of the 1950 Act under which the amount in dispute may be 'secured' does not give the court power to order a defendant to pay a sum into Court as a condition of obtaining a stay - *The Agrabele* [1979] 2 Ll. R. 117.

(v) Lack of the merits seems no bar to obtaining a stay.

Several aspects of the provisions mentioned above have occasioned controversy and are dealt with in turn below.

(c) Steps in the proceedings

A step in the proceedings generally means an application to the Court indicating an election to proceed with the action rather than to enforce the right to arbitrate. Thus, appearing before a master asking for leave to defend a summons for judgment under Order 14 – *Pitchers Ltd.* v *Plaza (Queensbury) Ltd.* [1940] 1 All ER 151, (CA) – and applying for security for costs – *Adams* v *Catley* (1892) 40 WR 570, (Div. Ct.) – have both been treated as steps, but not a request to the other side's solicitors (as distinct from taking out a summons) for further time to deliver a defence – *Brighton Marine Palace Ltd.* v *Woodhouse* [1893] 2 Ch. 486, nor contesting proceedings for an interlocutory injunction (since this is merely parrying a claim) – *Roussel-Uclaf* v *G. D. Searle & Co. Ltd.* [1978] 1 Ll. Rep. 225, nor even applying to strike out a defective statement of claim (because doing so did not affirm the correctness of proceeding through the Courts) – *Eagle Star Insurance Co. Ltd.* v *Yuval Ins. Co. Ltd.* [1978] 1 Ll. Rep. 357, (CA).

Filing an affidavit giving reasons for the master not to give judgment was fatal to an application for a stay; it mattered not that the solicitor did not know of the requirement that no step must be taken, or that he was unaware of the arbitration agreement and so did not realise the possibility of applying for a stay – *Turner and Goudy* v *McConnell* [1985] 2 All ER 34 (CA).

(d) Sufficient reasons not to refer

Among reasons which at one time or another have been treated as sufficient are undue delay in applying for the stay – *The Elizabeth H.* [1962] 1 Ll. Rep. 172; the fact that the arbitrator was suspected of bias – *Blackwell* v *Derby Corporation* (1911) 75 JP 129, (CA); and that arbitration would lead to a duplication of proceedings – *Taunton-Collins* v *Cromie* [1964] 1 WLR 633. In the *Blackwell* case the arbitrator was the corporation engineer and a main issue was whether the delay complained of was attributable to his unreasonable conduct as engineer. In the *Taunton-Collins* case, a plaintiff sued an architect for negligence and joined the builders. The builders whose contract included an arbitration clause were refused a stay mainly because of the risk that the arbitrator and court might reach different conclusions. Pearson LJ, thought the best practical solution would have been a tripartite arbitration had all the parties agreed.

In *Berkshire Senior Citizens Housing Association* v *McCarthy E. Fitt Ltd.* (1979) 15 BLR 27 CA the plaintiff sued the builders and the architect's executors under separate contracts both of which contained arbitration clauses. The first defendant applied for a stay but the second, because of the possibility of bringing in the engineers as third parties, did nothing. The judge, who ordered both actions to be stayed, was reversed on the ground that the second defendant could not be forced to arbitrate and the plaintiff could not be forced into a tripartite arbitration. To avoid the possibility of different findings the first defendant was refused his stay.

Inability to get legal aid in arbitration was, exceptionally, treated as such a reason in *Fakes* v *Taylor Woodrow Construction Ltd.* [1973] 1 QB 436 (CA) because the court considered that there was a reasonable probability of the plaintiff establishing that the defendants had contributed to his poverty.

The financial position of the plaintiff is, however, only one of the considerations to be taken into account, and a stay was granted in *Goodman* v *Winchester and Alton Railway* [1984] 3 All ER 594, CA, where the defendant company comprised a large number of railway enthusiasts and the arbitration clause had been inserted at the plaintiff's request.

(e) Existence of a dispute

The mere refusal to pay an acknowledged debt is probably not a 'dispute' for the purposes of S1(1) of the 1975 Act, but failure to respond to a claim might well be – see *Tradax Internacional SA* v *Cerrahogullari* cited at p. 36.

To entitle a defendant to a stay, his claim against the plaintiff must be capable in law of relating to the same matter. Thus, no stay will be granted in an action on a cheque on the ground that the goods paid for were defective, nor, in an action for freight, on the ground that the unseaworthiness of the ship necessitated a detour, the reason in each case being that even if the allegations were sustained they could not operate by way of set-off as distinct from counterclaim – see *Nova (Jersey) Knit* v *Kammgarn Spinnerei GmbH* [1977] 1 Ll. Rep. 463 (HL), and also, *A/S Gunnstein & Co K/S* v *Jensen Krebs (The Alfa Nord)* [1977] 2 Ll. Rep. 434, (CA), an action for freight with a counterclaim for delay.

Certain recent Court of Appeal decisions – such as *Associated Bulk Carriers Ltd.* v *Koch Shipping Inc. (The 'Fuohsan Maru')* [1978] 1 Ll. Rep. 24, and *S. L. Sethia Liners* v *Naviagro Maritime Corporation, (The 'Kostas Melas')* [1981] 1 Ll. Rep. 18 – see also *S. L. Sethia Liners Ltd.* v *State Trading Corp of India Ltd* [1986] 2 All ER 395 – have reinforced the view that a defence for Order 14 purposes and a dispute within the meaning of S1(1) of the 1975 Act are opposite sides of the same coin notwithstanding the differences in language. It follows that even in non-domestic arbitrations, and even where there is a valid exclusion clause, the courts have arrogated to themselves jurisdiction to decide issues which the parties expressly agreed should be determined by an arbitrator. This is arguably contrary to the 1958 Convention, and to Parliament's intention in the 1975 Act, and is the more unsatisfactory in that the court's decisions are based on affidavit evidence only. Nevertheless, it is doubtful whether the House of Lords would reverse the present run of authority though it is possible that the position is not quite as settled as Mustill & Boyd indicate – see p. 92 notes 5 and 6.

It follows from the foregoing that whether or not a dispute exists may cause much controversy and, as recently happened in an unreported case, experienced Counsel may argue long hours about matters on which they are wholly at variance only to find in the end that there was no dispute between them after all.

6. Arbitration as a condition precedent to litigation – *Scott* v *Avery* clauses

Parties who wish to render it virtually certain that any dispute will be settled by arbitration may insert what is called a *Scott* v *Avery* clause in the arbitration agreement. A typical example is: 'The making of an arbitral award shall be a condition precedent to the commencement of any action at law'. Another form of the clause limits the amount a party can recover to the amount which an arbitrator awards. The bar to litigation may be a partial one, for example, as to quantum only and not as to liability; it will not operate at all if it can be treated as waived.

S25(4) of the 1950 Act recognises the validity of such a clause and in fact gives the Court power to strike it out if it orders that the arbitration agreement is to cease to have effect as regards a particular dispute, whether the order is made under that Section or any other enactment (e.g. S24 or, it is submitted[4], Ss13(3) or 23(1) of the same Act provided the terms of S25(2) are satisfied).

7. Time limitations – *Atlantic Shipping* or *Centrocon* clauses

A *Scott* v *Avery* clause may be strengthened by introducing a time limit, for example, by providing that any claim under the contract fails if some action such as making a claim, appointing an arbitrator or commencing arbitration proceedings is not taken within a specified time. This is often referred to as an *Atlantic Shipping* or *Centrocon* clause, a typical wording being as follows: 'Any claim hereunder must be in writing and the claimant's arbitrator appointed within three months of final discharge of the ship failing which such claim shall be deemed to be waived'.

This could be a rather harsh rule and S27 of the 1950 Act gives the court power to extend the time limit in certain specified cases where its enforcement would cause undue hardship. The parties cannot exclude it, for example, by providing that time is to be of the essence – see *Pittalis* v *Sherefettin* [1986] 2 All ER 227, at 233 C.

The cases, which S27 specifies are those where a period is defined within which a party must appoint an arbitrator, give notice to appoint one, or take some other step to commence arbitration proceedings. Thus, a clause merely stating that a claim must be made within a given time is not covered – see *Mariana Islands Steamship Corp* v *Marimpex Mineraloelhandels GmbH The Medusa* [1986] 2 Ll. Rep. 328 (CA). However, in *Jedranska Slobodna Plovidba* v *Oleagine SA. The Luka Botic* [1983] 3 All ER 602, (CA) where a claim had to be made and an arbitrator appointed within three months of the discharge of a ship it was held that the time for making the claim as well as for appointing the arbitrator could be extended since the two steps were so bound together that the claim was a step in commencing the proceedings.

Hardship is 'undue' in this context if it is out of proportion to any fault of the claimant's – see *Liberian Shipping Corporation* v *A. King & Sons Ltd.* [1967] 1 All ER 934 (CA), where the claimant was only nine days late and the penalty would have been £33,000.

The criteria which the courts should use in deciding whether or not to extend the time have now been laid down in *Libra Shipping and Trading Corporation Ltd.* v *Northern Sales Ltd. The Aspen Trader* [1981] 1 Ll. Rep. 273 (CA), and recently re-affirmed in *Graham H. Davies (UK) Ltd.* v *Marc Rich & Co. Ltd.* [1985] 2 Ll. Rep. 423 (CA). They are as follows:
   The length of the delay;
   The amount at stake;
   Whether the delay was the applicant's fault and if so the degree of fault;
   Whether the applicant was misled by the respondents;
   Whether the delay has caused prejudice to the respondents.

S27 does not, however, apply to statutory arbitrations – see S31 *ibid*. Nor can it be used to override normal limitation periods.

If failure to observe a time limit bars the claim itself, the arbitrator should make his award accordingly as soon as the point is taken. If, however, it merely bars the right to arbitrate the arbitrator has no jurisdiction to deal with it but the claimant can bring an action.

Where the applicable arbitration rules – such as those of GAFTA, Rule 2:7 – give the arbitrators power to extend the term, there has been judicial disagreement as to whether their failure to exercise it shuts out the court. In the latest case on the subject *European Grain & Shipping Ltd.* v *Dansk Landbrugs Grovvareslskab* [1986] 1 Ll. Rep. 163, Leggatt, J, held that he did have the discretion (though in fact he refused to exercise it).

In *Tradax Internacional SA* v *Cerrahogullari TAS The M. Eregli* [1981] 3 All ER 344 an amended *Centrocon* clause required a complainant to appoint an arbitrator within nine months of final discharge. The plaintiff made various claims within that period without any response from the defendants and finally made an appointment one year out of time. The defendants ultimately admitted the claim but said it was now barred. It was held that failure to respond without admitting the claims did not mean that there was no dispute; consequently, an appointment ought to have been made. However, the court extended the period under S27, this being no hardship to the defendants who had no merits.

It was held in *The Himmerland* [1965] 2 Ll. Rep. 353, that where a *Centrocon* clause required an arbitrator to be appointed within three months a claim made after that period was barred even though the claimant did not become aware that he had a claim until after expiry of the three months. S27 relief was refused because of a delay in claiming it.

8. Termination of the arbitration agreement

The general rule is that once an arbitration agreement is validly made it will continue in being notwithstanding termination of the contract which contained it. In other words, it normally has a life of its own and can govern the rights of parties even after the frustration or other determination of the main agreement – *Heyman* v *Darwins Ltd.* [1942] AC 356. It follows that although an arbitrator normally has no power to rule on the validity of the main agreement – see Chapter 1, pp. 4–5 – he may decide whether it still remains operative.

An arbitration agreement cannot be altered unilaterally and remains binding notwithstanding the death of a party, though in certain cases death may extinguish the right of action itself – see S2(1)(3) of the 1950 Act.

The court has power to order that an arbitration agreement is to cease to have effect where whether a party has been guilty of fraud is an issue, or, on the application of a party, where the authority of an arbitrator or umpire has been revoked with leave of the court or the court has removed an arbitral tribunal as a whole – see S24(2) and 25(2) *ibid*.

As in the case of other contracts an arbitration agreement can come to an end by frustration or, where there has been a repudiatory breach, at the election of the other party. Frustration occurs where some event arises, without the default of either party, which the parties had not foreseen or provided for and which either renders performance impossible or radically different from what the parties had contemplated – *Paal Wilson & Co.* v *Partenreederie Hannah Blumenthal* [1983] 1 All ER 34 (HL).

An arbitration agreement may also be terminated by abandonment or recission by mutual consent.

To prove abandonment, a party must show *either* that this was to be inferred from the parties conduct *or* that having been reasonably led to believe that that was the other's intention he has altered his position accordingly, indicating what he did or did not do on the basis of his assumption; whether the other party knows of it does not matter since the latter's actual state of mind is not relevant for this purpose.

Notes

1. S7(1)(e) of the 1979 Act applies the same definition, while S7(1) of the 1975 Act comes to the same thing.
2. These being administrative provisions this up-dating would in fact occur even though not expressed – *Bunge SA* v *Kruse* [1979] 1 Ll. Rep. 279, at p. 286.
3. Contrast S3(7) of the 1979 Act where whether an arbitration agreement is domestic depends on the status of the parties at the time of the *agreement* – see Chapter 8, p. 99.
4. Mustill and Boyd state the contrary (at p. 130).

4 The offices of arbitrator and umpire

1. Appointment by the parties

Where a contract provides for arbitration in the event of a dispute it normally lays down the method of choosing the arbitrator(s). Sometimes a specific person is mentioned, especially where the dispute has already arisen, but often nomination is in the hands of someone in an official position such as the President of a professional body or trade association, assuming the parties have not agreed on an appointment in the meantime.

A common alternative is a provision that each is to nominate one arbitrator with these two appointing either a third arbitrator or an umpire, or waiting unless they disagree and then appointing an umpire, distinctions which are of some importance – see below. If the appointment is of a third arbitrator it is arguable that the arbitration agreement could be frustrated by a party failing to appoint his original arbitrator on the ground that the wording of neither S7 nor 10(c) of the 1950 Act strictly applied.

If he is the appointee of one of the parties he should ensure that the other is notified of his appointment and acceptance and if nominated by a third person that both the parties have been notified of his nomination and acceptance. If the arbitration agreement provides for the parties to agree on the appointment he should ascertain that they do in fact agree and have been duly notified of his acceptance. His actual appointment should be in writing, signed by the appointor(s), and his acceptance and all the notifications mentioned above should also be in writing.

A nomination is not complete until it has been notified to the other party; nor is an appointment perfected until the arbitrator has notified the parties, or their agents, of his acceptance, but this may be before or after notification to the other side – *Bunge SA* v *Kruse* [1979] 1 Ll. R. 279, at p. 295.

Where the arbitration agreement provides that each party is to appoint an arbitrator S7 of the 1950 Act supplements the provisions applicable unless the agreement indicates to the contrary. Paragraph (a) gives an appointor another choice if his appointee refuses, dies, or is otherwise incapable of acting. Paragraph (b) states that if a party who has made his own appointment serves a notice on the other calling on him to appoint and the latter neglects to do so for seven clear days, the former may nominate his appointee sole arbitrator. However, the court has power to set such an appointment aside.

A notice to appoint within seven days accompanied by a notice that on default the arbitrator appointed by the other party would act as sole arbitrator was held bad on two counts – they were not 'clear' days and the appointment was premature because it was made before the other party was in default – *Bangladesh Ministry of Food* v *Bengal Liners Ltd.* [1986] 1 Ll. Rep. 167.

It seems from the Irish case of *Yeates* v *Caruth* [1895] 2 IR 146 (CA), that S7 has no application where both arbitrators are appointed by the arbitration agreement and not one by each.

The fact that arbitrators are appointed by the respective parties does not affect their obligation to act with impartiality. However, subject to the terms of the arbitration agreement, where each party has nominated an arbitrator and the arbitrators have appointed an umpire there is nothing wrong in the arbitrators thereafter acting as advocates for the parties who respectively nominated them. Indeed, this is commonly done to save costs, a practice which has been recognised by the Courts and also some sets of rules – see, for instance, paragraph 2(b) of the LMAA Terms 1987.

2. Umpires

As mentioned on page 27, unless the arbitration agreement expresses a contrary intention it is deemed to provide that:
(a) Where the reference is to two arbitrators they may appoint an umpire at any time and are to do so forthwith if they cannot agree;
(b) If the arbitrators notify either party or the umpire in writing that they cannot agree, the umpire may enter forthwith on the reference in lieu of the arbitrators – S8 of the 1950 Act amended by S6 of the 1979 Act.

If the appointment of the umpire is left to the arbitrators it should be by the deliberate choice of both acting together. Thus, it will be void if the arbitrators toss a coin as to which of them shall choose – *Jamieson* v *Binns and Dean* (1836) 4 A and E. 945, though the parties may validate the appointment by acquiescing provided they know all the relevant facts. However, if the arbitrators both agree that either of two persons would be suitable but cannot agree upon which they may validly decide by lot – *Neale* v *Ledger* (1812) 16 East 51.

Where an umpire has been appointed, and a party so applies, the court may at any time order him to enter on the reference as though he were sole arbitrator whatever the arbitration agreement may say – S8(3) of the 1950 Act.

An umpire may sit with the arbitrators and even ask questions for elucidation purposes, but he has no powers and may take no part in their deliberations unless and until they signify in writing either to him or to the parties that they are in disagreement. When that happens, however, he takes over and effectively acts as sole arbitrator. How he should deal with actual decisions by the arbitrators prior to the disagreement is doubtful; presumably he could not overturn an interim award.

The position is quite different where the two arbitrators have power to appoint a third *arbitrator*, when the three have an equal voice and sit together throughout. Further, S9 of the 1950 Act as amended by S6(2) of the 1979 Act provides that, unless the arbitration agreement indicates otherwise, in a reference to three arbitrators any two may make a binding award. It may be noted in passing that this otherwise useful provision is of no help where, as sometimes happens, there are three possible answers each of which is supported by one arbitrator.

In practice, where two arbitrators are appointed it is sometimes appropriate for them formally to 'disagree' at an early stage, appoint the umpire, and then retire from the case. Otherwise, they may disagree later and the case will have to begin again with a great waste of costs; alternatively, if the umpire has been sitting in to avoid this, there may be no disagreement in which event the costs of the umpire will have been thrown away.

Although an umpire is almost invariably one individual this is not necessarily so; for instance, the arbitration agreement might appoint the committee or officers of an association or society to be the umpire.

3. The number of arbitrators

The general practice in domestic arbitration is to have a single arbitrator, and in fact S6 of the 1950 Act deems a reference to be to one arbitrator unless the arbitration agreement otherwise provides.

A three-person tribunal provides a measure of insurance against a wholly unreasonable decision which may be all the wiser now that the 1979 Act and the *Nema* guidelines have so reduced the opportunities to appeal. It has long been more usual in London maritime cases. Indeed, in *Laertis Shipping*

Corp. v *Exportadora Española de Cementos Portland SA* [1982] 1 Ll. Rep. 613, in a charter-party arbitration it was held on the evidence of eminent marine arbitrators that an appointment in the 'customary manner' was of an arbitrator by each party and an umpire by the two arbitrators. In maritime cases generally, however, it is perhaps more common for the third person to be another arbitrator in which case, with a view to saving time and expense, his actual appointment is sometimes delayed, with the parties' agreement, until the end of the interlocutory period – or even beyond in a documents only arbitration unless disagreement arises in the meantime.

The rules of some bodies prescribe larger numbers. Thus, where a dispute is referred to the Refined Sugar Association its Council is directed to appoint not less than five persons from its panel of arbitrators.

Where more than one arbitrator is appointed, the authority of each must be co-extensive.

The main disadvantages of having more than one are delay and cost; it is often difficult enough to find days which suit the parties, their various advisers and just one arbitrator.

However, occasionally there are particular reasons for appointing more than one. Thus, where the matter involves technical expertise of a high order in two fields, it may be difficult to find an arbitrator sufficiently versed in both. A satisfactory tribunal in such a case might be an expert from each field together with a lawyer, especially one with some degree of technical knowledge.

A three-man tribunal is also usually preferable in international cases, particularly where one party comes from a common law jurisdiction and the other does not. With a single arbitrator coming from one type of jurisdiction, the party from the other is apt to think he is at a disadvantage, and the procedural differences are such that the disadvantage may be a very real one. Such a party may be at least partially reassured if someone from his type of jurisdiction forms part of the tribunal.

4. Appointment by the Court

The court has no general jurisdiction to appoint arbitrators or umpires but may do so either with the specific consent of both parties – *Tzortzis* v *Monark Line* [1968] 1 Ll. Rep. 337 – or under S10 or S25 of the 1950 Act on the application of one of the parties. A person appointed by the court under either of these sections has the same power to act as if he had been appointed by the parties in the normal way.

(a) S10 of the 1950 Act as amended by S6(3)(4) of the 1979 Act applies where the vacancy did not arise as a result of the court's own action. It may appoint an arbitrator or umpire where (i) the arbitration agreement provides for a single arbitrator, a dispute arises, and the parties cannot agree upon whom to appoint; (ii) an arbitrator or umpire refuses to act or is incapable of doing so and the vacancy is not filled – unless the arbitration agreement indicates it is not to be filled; (iii) the parties, or two arbitrators, are at liberty or required to appoint a third arbitrator, or umpire, and fail to do so; or (iv) a third party who is given the task of appointing an arbitrator or umpire refuses, or fails to appoint within the time specified, or within a reasonable time if no time is specified.

The procedure is that one party serves the other (or the arbitrators, or such third party, as the case may be) with written notice to appoint, or to concur in an appointment and, if there is no compliance within seven clear days, that party may apply to the court to make the appointment.

(b) S25 of the 1950 Act, on the other hand, applies where the High Court (or Court of Appeal)[1] created the vacancy by removing one or more members of the tribunal. Provided that at least one arbitrator is left, or on the removal of an umpire who has *not* entered on the reference, the court may simply replace him or them (on a party's application) – subsection (1).

However, the court has an alternative where it has removed either the sole arbitrator, all the arbitrators, or an umpire who *has* entered on the reference, or has given leave for the authority of an arbitrator(s) or umpire to be revoked[2]. Then, on the application of a party, the court may either appoint a sole arbitrator or order that the arbitration agreement is to cease to have effect in regard to the dispute in question – S25(2) *ibid*. In the latter case, it may override any *Scott* v *Avery* arrangement which might otherwise restrict or bar action in the courts by the parties – see S25(4) of the 1950 Act and p. 34 *ante*.

5. Judicial appointments

If the arbitration agreement provides for the reference to be to an official referee he must act, subject to any court order 'as to transfer or otherwise' – see S11 of the 1950 Act.

Where a dispute is of a commercial character a judge of the Commercial Court may accept appointment as an arbitrator or umpire, if he thinks fit and the Lord Chief Justice agrees – S4 of the Administration of Justice Act, 1970. Schedule 3 to that Act modifies the Arbitration Act 1950 in relation to judge-arbitrators, while S4(5) provides that any jurisdiction exercisable by the High Court in relation to arbitrators and umpires otherwise than under the 1950 Act shall be exercisable instead by the Court of Appeal.

The differences in procedure resulting from judicial appointments are happily summarised in Mustill and Boyd at pp. 238 and 232–4.

6. Powers exercisable in an arbitration

(a) General

Since arbitration in the sense used here is a consensual proceeding, the main powers given to the arbitrator (or umpire) are those conferred by the parties themselves, whether directly, or indirectly by their agreeing to the application of a particular set of procedural rules which affect his powers – see Chapter 5 para. 2(a).

In addition, he has certain powers at common law corresponding to those of a court's 'inherent jurisdiction', while further powers may be conferred by the custom of a trade. The former comprise those necessary for the proper discharge of his functions, for example, to make orders for discovery and to call for evidence to be given by affidavit, but not powers which are collateral such as to grant interim injunctions. Power in 'look-sniff' arbitrations to examine the subject-matter of the dispute in the absence of the parties is an example of a power conferred by custom.

These powers are or may be supplemented by S12 of the 1950 Act or S5 of the 1979 Act.

(b) S12(1)(3) Arbitration Act 1950

Subsections (1)(3) of S12 confer some miscellaneous powers on the arbitrator unless the arbitration agreement contains a 'contrary intention'. Though set out at some length, these amount to little more than that the arbitrator (or umpire) may take evidence on oath or affirmation, and that the parties and those claiming through them, unless entitled to object on legal grounds, must give evidence, produce documents, and do whatever else is required of them at the hearing. To this limited extent there is statutory encouragement for the proceedings to be inquisitorial.

With the exception just mentioned the arbitrator does not himself have power to summons witnesses. It is for the parties to initiate the necessary action by applying to the High Court to issue subpoenas compelling third parties in the United Kingdom to attend as witnesses. Such subpoenas are of three kinds, namely, a *subpoena ad testificandum* where a person's oral evidence is required, a *subpoena duces tecum* where, conduct money having been tendered, he is required to bring documents (which must be identified and be of such a nature that their production could be demanded in an action), and a *habeas corpus ad testificandum* where a person has to be brought from prison to give evidence – S12(4)(5).

(c) S12(6) Arbitration Act 1950

S12(6) of the 1950 Act gives the court power to make orders similar to those it could make in a High Court action in regard to a whole rag-bag of matters including security for costs, discovery, interrogatories, giving evidence by affidavit, examining witnesses on oath before an officer of the High Court, issuing commissions for the examination of witnesses outside the jurisdiction, the inspection, preservation, interim custody, detention and sale of property forming the subject matter of the reference, securing the amount in dispute, authorizing anyone to enter upon land or premises in the possession of a party to obtain information or evidence (including taking samples and carrying out experiments), making interim injunctions and appointing a receiver.

The powers given by S12(6) are expressed to be without prejudice to those already vested in the arbitrator (or umpire) and include some he has at common law. The reason for this apparent duplication may be to enable the court to override the expressed wishes of the parties in certain circumstances; certainly the usual saving for cases where the arbitration agreement expresses a contrary intention is absent. However that may be, the power to make the orders is discretionary and it is not the normal practice of the court to confer powers on an arbitrator which he is reluctant to possess.

(d) S5 Arbitration Act 1979

S5 of the 1979 Act enables the court to re-inforce the arbitrator's powers in relation to interlocutory matters upon an application by the arbitrator (or umpire) or, which is more likely in practice, by a party. Its apparent intention is to reinforce the tribunal's power to proceed with the reference in the face of obstructive tactics. Subsection (1) sets the scene – non-compliance with an arbitrator's order within the time specified, or within a reasonable time if no period is specified, followed by an application to the court by the arbitrator or a party. An order under S5(2) will give the arbitrator power, subject to such conditions as may be laid down, to continue in the way a judge might do in comparable circumstances.

Unlike S12(6) of the 1950 Act which requires a court order, this rather oddly worded[3] provision extends the arbitrator's powers to issue directions himself within the scope of the authority given him, but in practice may not greatly extend the powers which the arbitrator (or umpire) already has at common law. Thus, as mentioned at p. 58, he does not need the court's help to issue a peremptory order or, ultimately, to proceed *ex parte*, and it is not clear that he can be given one important power he otherwise certainly lacks, namely, to dismiss a claim for want of prosecution. A S5 order can, however, provide greater flexibility to the exercise of certain powers and perhaps avoid having to apply to the court more than to the minimum extent.

A judge-arbitrator can make a S5 order himself.

7. Remuneration

It is common, and certainly desirable, for the basis of the arbitrator's remuneration to be agreed at or before the time of his appointment or, at the latest, at the preliminary meeting, and for it to be confirmed in writing. In a simple case, this may be a fixed amount; otherwise, it is usually so much per day or half-day for hearings, and so much per hour otherwise. It should be noted that if a fee is agreed in writing and a party later considers it excessive he cannot adopt the procedure of paying the amount demanded into court and forcing the arbitrators to go through taxation – see S19 of the 1950 Act and page 86.

An arbitrator (or umpire) has power to tax his remuneration. He may also exercise a lien on the award for it. See Chapter 7 at pages 86-7.

It may sometimes be wise for an arbitrator to stipulate for a payment on account at the commencement of the arbitration, and perhaps for further sums as the arbitration proceeds, especially where the parties are not represented by solicitors.

In the absence of an agreement, it will be presumed, at least in the case of a commercial arbitration, that an arbitrator (or umpire) is entitled to be paid a 'reasonable amount' – see *Brown* v *Llandovery Terra Cotta Co. Ltd.* (1909) 25 TLR 625, where an arbitrator appointed by a losing party was held entitled to recover his fee from the successful party since there is an implied promise by parties to a submission jointly to pay the arbitrator (and umpire) for acting.

Arbitrators have no inherent power to call for security for their costs though some of the published sets of rules make provision for it, but an application may be made to the court for an order under S12(6)(a) of the 1950 Act. In considering such an application, the court will take into account such matters as that one or both parties are foreign corporations or have no assets within the jurisdiction, and also, where both parties are foreign, whether they have shown the intention to apply English law to their commercial relationship or the connection with England is merely that of a convenient venue for the arbitration – see *Bani and Havbulk v Korea Shipbuilding and Engineering Corp*. FT Law Reports 10 July 1987. At a later stage security may be required for actual outgoings, such as the cost of a shorthand writer.

Where the parties settle before the award, the arbitrator can normally recover his fees – or a proper proportion of any lump sum agreed – but not damages for loss of the opportunity to earn a greater sum. However, fees probably cannot be recovered where the termination does not stem from actions of the parties, for example, the arbitrator dies, and *a fortiori* if he is removed or his award set aside by the court, or if he knew he lacked qualifications required by the arbitration agreement.

An arbitrator who is removed by the court for failing to use all reasonable despatch in entering on and proceeding with the reference and making an award cannot recover any remuneration for his services. This would apply to two arbitrators who disagreed and were unreasonably tardy in notifying the parties and the umpire – see S13(3) of the 1950 Act.

If an umpire attends the proceedings but no fee is agreed he will *prima facie* be entitled to remuneration on a *quantum meruit* basis even though he never enters upon the arbitration because the arbitrators do not disagree.

8. Removal of an arbitrator or umpire

The parties can always remove an arbitrator (or umpire) by agreement and will do so automatically if they settle the case.

Where only one party wishes the removal he must apply to the Court which however does not lightly exercise its discretion to order it. Circumstances in which it has been exercised include the following:
(a) where the arbitrator had an undisclosed interest in the matter in dispute;
(b) where he subsequently acquired an interest of sufficient significance to affect his judgment;
(c) where subsequent events suggested he might have developed a bias, whether for or against a party;

(d) where he had misconducted himself or the proceedings – S23(1) of the 1950 Act. 'Misconduct' here does not necessarily mean something morally wrong but may, for instance, include a breach of correct procedure. See, per Donaldson, J., in *Thomas Borthwick (Glasgow) Ltd. v Faure Fairclough Ltd.* [1968] 1 Ll. Rep.16, at p. 29, a case where a tribunal was held to have acted unfairly. Another illustration is *Walford, Baker & Co. v Macfie and Sons* (1915) 84 LJKB 2221, where the arbitrator acted on a term which appeared not in the relevant contract but in an earlier one between the parties which he had wrongly admitted in evidence.

(e) Where he has failed to use all reasonable despatch – see p. 46, *ante*.

(f) As mentioned, at p. 39, the High Court may set aside an appointment made under S7 of the 1950 Act.

An arbitrator may be removed or his award set aside if reasonable suspicion of bias arises as in the case of a planning inspector who entered into a long conversation with the chairman of a planning authority immediately on the conclusion of an enquiry – see *Simmons* v *Sec. of State for the Environment* [1985] JPL 253. The position may be the same if he receives documents from one party without ensuring that the other receives copies.

9. Revocation of the authority of an arbitrator or umpire

S1 of the 1950 Act provides that the authority of an arbitrator or umpire appointed by or by virtue of an arbitration agreement is irrevocable[4] except with leave of the court unless the agreement expresses a contrary intention.

To induce the court to act a party must make out a very strong case. Instances where it has given leave are where the arbitrator exceeded his jurisdiction or was guilty of misconduct such as would provide grounds for setting aside his award in due course, or where he lacked the qualifications for appointment contained in the arbitration agreement.

Once a dispute has arisen, an application for an injunction to restrain an arbitration from proceeding, or for leave to revoke the authority of an arbitrator named or designated in an arbitration agreement because he is or may not be impartial, will not be refused on the ground that the applicant knew or should have known this when he entered into the agreement – S24(1) of the 1950 Act.

Where fraud by any party is an issue in a dispute, the High Court has power to order the arbitration agreement to cease to have effect and to revoke the authority of an arbitrator or umpire appointed under it so far as may be necessary to enable the court to determine that issue – S24(2) of the 1950 Act. This is not so, however, in the case of a non-domestic arbitration agreement where there is an exclusion agreement – as to which, see p. 99 – S3(3) of the 1979 Act.

Where the High Court has taken action under S24 it may refuse a stay and has the powers of appointment, etc. mentioned in section 4 *ante*, at p. 42.

In addition to its statutory powers, the court has inherent jurisdiction to prevent an arbitrator from acting if it considers that he is unfit to do so or has been improperly appointed.

Death of a party does not revoke the authority of an arbitrator whether or not appointed by the deceased – see S2(2) of the 1950 Act.

10. Retirement and resignation

An arbitrator has no authority to retire unless entitled to do so under the arbitration agreement, which is highly unlikely, but there would be nothing to prevent the parties agreeing to accede to an arbitrator's wishes to retire. It is very doubtful whether an arbitrator appointed to take over the arbitration in his place would be entitled to treat even modest charges by his predecessor as constituting part of the costs without the agreement of the parties.

In certain circumstances an arbitrator might properly resign at the request of a party. One instance might be where a material without prejudice offer had been disclosed to him or where changed circumstances resulted in his acquiring a substantial interest in the outcome of the arbitration, e.g. as a result of a merger. In such cases he would presumably be entitled to a fee on a *quantum meruit* basis unless personally responsible for the need to resign.

There may also be circumstances in which an arbitrator should apply to the court to be relieved from acting further – see, for instance, *Abu Dhabi Gas Liquefaction Co. Ltd.* v *Eastern Bechtel Corp.* [1982] 2 Ll. Rep. 425, at p. 427.

The rules applicable to an arbitration may permit resignation. Under Rule 9 of the London Bar Arbitration Scheme an arbitrator must resign if unable to offer a hearing date within a reasonable time unless the parties request him not to do so.

Notes

1. The references to the Court of Appeal were introduced by the Administration of Justice Act 1970, S4(4), to deal with cases where a judge of the Commercial Court is appointed arbitrator or umpire – see 'Judicial appointments' at pp. 42–3.
2. The removal of arbitrators and umpires and the revocation of their authority, are dealt with in sections 8 and 9 at pp. 46 and 47, respectively.
3. Thus the phrase 'in default of appearance' has no application to arbitration as distinct from litigation; just possibly it is intended to mean simply failing to turn up at a hearing.
4. At Common Law the appointment could have been revoked at any time before the award.

5 Procedure before the Hearing

1. Initially

When a person is approached to act as an arbitrator (or umpire), he should first ascertain that he is not disqualified by any evident conflict of interest or by any provision of the arbitration agreement, that the terms upon which he is prepared to act are acceptable to the parties and that he has been properly appointed – see p. 38–9.

Once satisfied that his appointment is in order he should examine the material he will have been sent initially in order to judge, so far as he may, the nature of the differences which have arisen and make sure that they fall within his terms of reference. If there is any doubt on the point, he should invite the parties to resolve it by making an appropriate amendment to the arbitration agreement.

One important procedural matter will inevitably be dealt with very early on, namely, whether the arbitration is to be 'administered', that is to say, whether the administration is to be undertaken by some outside body, perhaps the one whose President nominated the arbitrator. If there is a choice, it is better for the arbitrator to undertake the administration himself unless he is inexperienced or lacks the necessary resources.

2. The preliminary meeting

In all but very simple cases and those where costs must be minimal, it is generally best for the arbitrator to begin by calling the parties, with or without their immediate advisors, if any, to a meeting. It is desirable that the parties themselves should attend, or an executive of standing in the case of a limited company, since it may help to reconcile the advisors to a shorter time-table than they would otherwise have proposed – and make it more difficult to blame the arbitrator for any subsequent delays. The attendance of Counsel, on the other hand, is not generally necessary unless important issues are expected to arise at the interlocutory stage. For the different procedure often adopted in London maritime cases, see section 4 *post*.

If one side is bringing Counsel, the other side should be informed beforehand to avoid its being taken by surpise. Indeed in practice it is often helpful if both sides give the arbitrator a list, copy to each other, of all those expected to attend.

In an important or complex case it may well be wise to provide the parties with an agenda in advance so that they may come prepared to give their views on the main issues which will be raised.

In a technical matter, it may save a great deal of time to have a discussion with the proposed expert witnesses, or the parties' own experts, before the pleadings and either at, or even before, the preliminary meeting so that the issues are clearly defined and irrelevancies eliminated early on. Indeed, in complex cases the experts may need to clarify the issues for themselves so that they can address the correct points in their reports.

The preliminary meeting – or perhaps the first one, as there may be others – corresponds to a summons for directions in a High Court action. It is usually the arbitrator's first real opportunity to impress his personality upon the parties and make it clear to them, and to the solicitors if solicitors are instructed, that he will be streamlining the procedure as far as possible and that he will not willingly allow the case to meander on as so often happens in litigation.

The meeting will also give the arbitrator the chance to find out more about the actual issues and the background to the case. He may even have the opportunity to make some progress towards a settlement, or settlement in part, but should remember that while ways of reducing the scope of an old dispute may occur readily enough to a newcomer, expressing them may demolish long-cherished and deeply-felt notions. In fact it requires considerable skill and tact to pinpoint the important issues at an early stage and identify aspects which are susceptible to compromise without upsetting one or other of the parties.

(a) Procedural rules and time-table

As has already been pointed out the Arbitration Acts do not contain a procedural code which is in any way complete and it is often thought desirable to supplement the arbitrator's statutory powers. Furthermore it is convenient to operate by reference to a set of rules of which everyone has a copy readily available. If none are applicable already, the preliminary meeting is probably the arbitrator's last chance to canvass the parties' acceptance of the one which, with or without amendment, he considers most appropriate.

There are a number of sources of published arbitration rules; they may be grouped as follows:

(i) *International arbitrations.* Reference to some of these rules will be found in Chapter 9 – pp. 106–7.

(ii) *Arbitral organisations.* The rules most likely to be of general use in England are those of the Chartered Institute of Arbitrators – either the version intended for domestic use or the one designed for use in international arbitrations, as appropriate.

Another admirable set of rules is that of the London Maritime Arbitrators' Association, 1987, to which references are made elsewhere in this book.

(iii) *Rules of professional and trade associations.* Many associations have their own arbitral rules for use primarily by their members, among them the Royal Institution of British Architects and the Grain and Food Trade Association.

(iv) *Statutory Arbitrations.* Statutes which provide for 'arbitration' frequently contain their own set of rules usually based on, or supplementary to, those in the Arbitration Acts and, in particular, the 1950 Act. See Chapter 1 section 11.

Illustrations of statutes with their own arbitration rules are the Agricultural Holdings Act 1986, S84(1) of which provides that the 1950 Act is not to apply, and the Building Societies Act 1962.

If a published set of rules is adopted it will no doubt include a timetable for the rendition of pleadings but the arbitrator, having heard what the parties have to say, should not hesitate to speed it up or slow it down if the rules permit. In general, he should steer between giving the impression that a leisurely tempo will satisfy him and creating a genuine difficulty in the preparation of either party's case. It should also be borne in mind where solicitors are involved, that one or both of them at some stage – perhaps every stage – is likely to ask for and be granted an extension of time by the other side. The arbitrator should discourage this practice so far as it is reasonable to do so but can do nothing if the parties themselves countenance it.

(b) Claimant and respondent

It is not usually difficult for the arbitrator to determine which party is the claimant (or plaintiff) and which the respondent (or defendant) but where there are claims and counter-claims the position can be rather complicated and the general rule, namely, that the plaintiff is the party who would lose if no proceedings were brought, may become difficult to apply. In practice, it is usually right to assume that the party who initiated the proceedings is the claimant but if the arbitrator is in any doubt he must have it out at the preliminary meeting and, having listened to both sides, make a decision. The question is of considerable importance procedurally because the claimant opens the proceedings at the hearing and normally has a right of reply, and also because the onus, at least initially, is on him to prove his case.

(c) Clarifying the issues

If normal court procedure is followed, clarification of the issues will mainly be effected by means of pleadings the most important of which are:

(a) particulars of the claim;

(b) the defence, together with any counterclaim;

(c) the claimant's reply and his defence to any counter-claim; and

(d) the reply to the counterclaim, if any.

They are frequently referred to as 'points of claim', etc.

A 'pleading' is a document in which a party sets out his case for the benefit of the tribunal and the other side. 'Every pleading must contain, and contain only, a statement in a summary form of the facts on which the party pleading relies for his claim or defence, as the case may be, but not the evidence by which those facts are to be proved, and the statement must be as brief as the nature of the case admits'. RSC Order 18 r.7(1).

Pleadings are not necessarily the best, let alone the only, way to clarify the issues. Their main disadvantages are that they encourage a negative attitude by over emphasis on the denial of the other side's case, and frequently do not pinpoint issues of law which constitute important points of difference. They may even be used to conceal the real issues.

An alternative sometimes used is a Statement of Case in which the claimant sets out as shortly as may be the material facts together with the documents and other evidence on which he relies and any proposed contentions of law, while the respondent counters with a Statement indicating which parts of it he accepts and which he disputes, adding, if appropriate, his own Statement of Case. Discovery may then be confined, in continental fashion, to other relevant documents whose existence is known or suspected.

Where there are many individual items in issue it may be convenient for the plaintiff to set them out in a *Scott Schedule* in which the Respondent indicates against each item the extent of his agreement, if any, together with any comments, leaving a spare column for the arbitrator to use. See an illustration at p. 114.

(d) Discovery and inspection of documents

Where an order for discovery is made each party is obliged to provide the other with a list of *all* documents relevant to the issues in dispute which are or at any time have been in his possession or under his control. Privilege may, however, be claimed in regard to certain classes of document, for example, instructions to, and the advice of, Counsel, communications between solicitor and client, documents prepared with a view to litigation, documents showing title to a party's own property and incriminating documents. Documents other than those which are privileged are open to inspection by the other party.

Inspection normally takes place at the offices of the solicitor of the party whose documents are being discovered shortly after the parties have exchanged their lists.

Most standard sets of rules give the arbitrator the right to order discovery, but if he lacks the necessary power and one of the parties is against giving it to him the other may apply to the High Court under S12(6)(b) of the 1950 Act.

S12(1) of the 1950 Act puts a similar point rather differently. Unless a contrary intention is expressed, an arbitration agreement is deemed to provide that the parties and all persons claiming through them are to produce before the arbitrator or umpire all documents within their possession or power which he calls for other than those whose production could not be compelled in an action (e.g. privileged documents).

An arbitrator has no power himself to order discovery against the Crown but, on application, the High Court will make the order, if appropriate. The Crown may claim privilege on the ground that disclosure would adversely affect the public interest, but such claims sometimes have regard mainly to the convenience of the civil servants concerned instead.

The Court may require to see documents in respect of which privilege is claimed with a view to determining whether the injury to the public interest is sufficiently substantial to justify possible frustration of the administration of justice – *Conway* v *Rimmer* [1968] AC 910 – and may be less sympathetic than usual to arguments based on confidentially since arbitration proceedings are held in private.

Discovery takes place almost automatically in litigation but this should not be the case in arbitration. It is true that discovery may clarify the issues by indicating the direction of the other party's case and may lead to a settlement through disclosure of its strength. Nevertheless, the procedure is often as expensive as it is tedious and if he orders it the arbitrator should seek to limit its scope as far as possible, especially if he suspects that the main motive is delay, or 'fishing', i.e. rummaging amongst the other side's papers with a view to bringing to light some claim or defence not previously provable or known to exist. If asked to make an order, the arbitrator should question both sides with a view to establishing the most appropriate limitations, e.g. by reference to dates, particular topics or issues, and/or correspondence with other than specified third parties. Where a case turns on expert evidence, little discovery may be needed. In fact, whether or not the case is a technical one the limitation of discovery depends very much on the extent to which it proves possible to define the issues at an early stage.

Particular care should be taken with ordering discovery when one of the parties comes from a civil law jurisdiction since, in general, it is a common law concept only. It is worthy of note that the exceedingly wide scope of discovery in US litigation is much reduced under the procedures of the American Arbitration Association.

(e) Agreed bundles

The arbitrator should normally ask the parties to prepare an agreed bundle(s) of correspondence and other documents and to deliver a copy to him well before the hearing. This saves time and minimises the shuffling of papers at hearings. However, especially in a case which is in any way complicated, he should make it clear that the bundles are to be confined to correspondence and other documents upon which one or other of the parties relies as being strictly relevant to the issues between them or which form part of the necessary background, and should normally seek to impose some limit upon them at least as restrictive as that mentioned in relation to discovery on p. 54. Otherwise the arbitrator may well be overwhelmed, and the hearing held up, by xeroxed copies of every document in sight.

This intractable problem may be minimised by directing the separate bundling of all documents upon which either party thinks the arbitrator should study before the hearing. Adoption of this suggestion should enable the opening address on behalf of the claimant to be considerably shortened.

(f) Inspection and protection of property

Subject to any provision to the contrary in the arbitration agreement, the arbitrator has inherent power to inspect any property or thing in the possession or under the control of a party which is relevant to the issues before him. The court has power to authorise (a) entry to any land or building in the possession of a party, and (b) any sample-taking or experimentation, etc., which may be expedient in order to obtain full information or evidence – see S12(6)(g) of the 1950 Act.

Inspections should take place in the presence of the parties or their advisers. They are often made after the hearing has begun, or at its conclusion when the arbitrator knowns precisely what to look for, but that is a matter entirely for him having considered the circumstances generally.

The court may also make orders for the preservation, interim custody or sale of goods which are the subject-matter of the reference under S12(6)(e).

(g) Other matters requiring decision

The arbitrator should take the opportunity to deal with any other procedural points which arise. Thus, when expert evidence is likely to be needed he should place some limit on the witnesses on each side, and cause them to exchange proofs of evidence and agree whatever they can.

There may also be problems with important witnesses – age, poor health, living abroad, and so on – and he may have to decide that part of the hearing be adjourned to a different venue or that evidence be taken on commission.

Whether a shorthand writer should be used may also be ventilated at the preliminary meeting; an arbitrator should take his own notes and has no inherent power to employ one and charge the cost to the parties. When the parties consider it would be convenient to have a shorthand writer present a convenient course is for the arbitrator to agree provided he is supplied with a copy of the transcript. He can in due course state in his award who is to pay the cost.

(h) Reasons and exclusion agreements

There are two other matters which can usefully be raised at the preliminary meeting though the parties may consider it premature to make an actual decision so early. The first is whether the parties anticipate entering into an exclusion agreement shutting out any appeal to the Courts, and the second whether a reasoned award will be required. If the answer to the latter is negative, an exclusion agreement may in practice be unnecessary.

These matters are dealt with in greater detail in Chapter 8. Suffice it here to say that it is helpful if the parties agree that should neither ask for a reasoned award before the hearing neither will do so later. This is because should either party take advantage of the full period allowed by S1(6)(a) of the 1979 Act for calling for reasons, namely, any time up to the issue of the award, a considerable amount of inconvenience, delay and additional cost is likely to result. Besides, if the arbitrator knows that reasons will not be required his note of the evidence and argument may possibly be a little shorter.

(i) Preparations for the hearing

It is usually a good idea at the preliminary meeting to fix a provisional date for the hearing and an estimate of how long it will last with a view to obviating so far as possible a request later on for an adjournment on the ground that a party, adviser or witness is otherwise engaged on the day(s) in question. Some allowance should be made for unexpected delays and it must be accepted that in practice the date frequently has to be put back while initial estimates of the length of the hearing often need drastic revision.

It is a good plan when substantial questions of law are in issue to call on the parties to produce their arguments in writing, say, one week before the hearing. Any decided cases or textbooks upon which reliance is placed should be indicated and photostat copies of relevant passages made readily available. Alternatives also directed to the saving of time at a hearing are the submission of a summary of issues and the holding of a pre-hearing review – see page 59, paras (e) and (g).

The prime responsibility for making physical arrangements for the hearing – an appropriate venue, etc. – falls on the claimant though an experienced arbitrator can help materially with suggestions. Clearly every effort should be made to meet the convenience of all concerned.

(j) Directions

The arbitrator should be careful to listen not only to what may be said by or on behalf of the parties on all interlocutory matters but to indicate what he is inclined to decide so as to help the parties, especially reticent ones, to make suitable representations. However, subject to any joint requests by the parties and to any procedural rules which it is not within his power to modify, the decisions are for him.

Soon after conclusion of the preliminary hearing the arbitrator should make an order incorporating what he has decided on all relevant matters and despatch a copy to each of the parties. An illustration of such an order will be found, at p. 112.

3. Later procedural matters

Matters requring the arbitrator's decision frequently arise after the preliminary meeting. It may be no more than a request by one side, opposed by the other, for extra time for submitting a pleading. If a reasonable excuse is given the request should be granted for at least part of the additional time unless it would involve some injustice to do so. If the request seems unreasonable or is made a second time the arbitrator may summon a meeting to hear both sides and, if a party is at fault, discourage him by awarding the costs of the meeting to the other side in any event.

(a) If a party fails to comply with directions and remains recalcitrant the arbitrator may issue a *peremptory notice* and, if that is ignored, at the request of the other side threaten to proceed with the hearing at a given time and place, *ex parte* if necessary. If this does not have the required effect, he should carry the threat out. See an illustration on p. 115.

Among matters which may call for careful consideration by the arbitrator or umpire during the interlocutory period are requests for *further and better particulars* and *notices to admit*, and for the administration of *interrogatories*.

(b) The purpose of the Particulars, which are the most common, is or should be, to clarify the issues where a pleading is insufficiently explicit. Requests are directed in the first place by one party to the other and the arbitrator will only be approached if they are refused. Before making an order – for which no specific authority is required – he may have to consider what weight to give to such arguments as that the Particulars cannot be given until after discovery or that a defence cannot be delivered until they have been supplied. He should also bear in mind that applications for Particulars are frequently made merely to embarrass an opponent or cause delay.

The arbitrator should only make the necessary direction if he concludes that the Particulars are needed to help the party asking for them to prepare his case. If he does, he must decide how long to give for their delivery and whether consequential alterations to the provisional time-table will be needed. According to his view of the circumstances, he may announce that he will award the costs to one side or the other, or that they will be dealt with at the end of the arbitration with the other costs.

If Particulars are not given the arbitrator may issue a *peremptory notice* warning the party in default that unless they are delivered within a specified time he may be debarred from relying on the point involved – see per Lord Diplock in *Bremer Vulkan Schiffbau und Maschinenfabrik* v *S. India Shipping Corp.* [1984] AC909 at pp. 986 and 987.

(c) *Notices to admit* facts or documents. The main purpose is to avoid having to prove something which is normally not controversial but formal proof of which may be very costly. Unreasonable failure to admit may be penalised by requiring the offending party to pay the costs of proof even if he succeeds on all the main issues.

(d) *Interrogatories* take the form of written questions which one party addresses to the other calling on him to make some admission which is helpful to the former's case, or which limits some aspects of the latter's. They also *should* serve to clarify the issues and save time, but again their purpose is often mainly tactical.

Parties are rarely entitled to interrogatories; they are not within the inherent powers of an arbitrator so if any applicable procedural rules give no help a High Court order under S12(6)(b) of the 1950 Act will be required.

(e) In a complicated case it is a good plan to ask the parties to produce a joint *summary of issues* shortly before the hearing, alternatively, a summary so far as agreed supplemented by a note of any additional matters which either side considers is in issue. This has many advantages. In particular, it can concentrate the minds of the arbitrator and both advocates on what are the essential arguments, documents and evidence, enabling opening remarks to be reduced and minor issues to be discarded.

(f) Applications to amend a pleading or statement of case should normally be granted unless to do so would cause injustice to the other side, which is not very likely, or raise issues which are outside the arbitrator's jurisdiction, which is even more unlikely. As a general rule the party asking for the amendment should pay the resulting costs in any event, unless he was not to blame.

(g) Finally, in cases which are likely to be long and complicated it is usually an advantage to hold a *pre-hearing review* attended by the advocates and solicitors on each side. Its main purposes are to consider the extent to which it has been practicable to comply with previous directions, what procedures can be adopted with a view to reducing costs, and whether any issues upon which neither party seriously relies can be eliminated.

4. Pre-trial procedure in London maritime cases

Preliminary meetings in London maritime cases have hitherto been relatively uncommon and, where held, often take place after discovery. This is because the practitioners – that is to say, the arbitrators, those who practice before them, and the instructing solicitors – comprise a comparatively small and compact band amongst whom a well-recognised practice has evolved. In fact, the parties' representatives are often asked to meet first to agree their own directions so far as they can thereby saving the arbitrators' fees and time.

However, where cases are expected to last more than five days, or special circumstances obtain, the Association's rules, the LMAA Terms (1987), now provide for representatives of the parties to review the progress of the case before requesting a hearing date with a view to reaching agreement on the conduct of the hearing and making any further necessary preparations. An informal meeting with the tribunal will then be sought for any final directions. The Fourth Schedule to those Terms contains a checklist of matters appropriate for consideration at this meeting which most arbitrators would find it helpful to scan through at some time during the interlocutory period.

It is worthy of note that in maritime cases the tribunal, the parties, their representatives and the witnesses are wont to lunch together, a practice which has more than one advantage.

5. Termination of the arbitration prior to the hearing

Arbitration proceedings are frequently brought to an end by the parties settling, whether with or without the arbitrator's help. This may simply take the form of a joint decision to end the arbitration on some agreed basis, for example, that one should pay the other a sum of money and/or perhaps withdraw certain allegations, with the costs being dealt with in some specified way. Such a decision terminates the dispute and abrogates the arbitration.

Alternatively, the parties may call upon the arbitrator to make an award by consent. This is normally preferable because it facilitates enforcement of the agreed terms should this prove necessary. It also brings into effect the doctrine of *res judicata* thereby preventing either side in the future from bringing proceedings arising out of the same facts, or from raising the same or new arguments on similar facts.

It may be possible to bring arbitration proceedings to an end, or to prevent their being commenced, by applying for an injunction or a declaration, or by bringing an action in the Courts. The last, of course, is only effective if the other party is not successful in applying for a stay. See Chapter 3.

6 The hearing, alternatives, and special cases

The arbitrator's control of procedure

Subject to any rules which bind him, and to any joint directions by the parties, the arbitrator is master of the proceedings before him and generally speaking may conduct them in whatever manner he thinks fit provided he gives each party a proper opportunity to present his case and breaches none of the other principles of natural justice - see *Abu Dhabi Gas Liquefaction Ltd.* v *Eastern Bechtel Corp.* [1982] 2 Ll. Rep. 425 (CA). In that case, Watkins, LJ, said that the court could not impose conditions when appointing an arbitrator and had 'no power to direct an arbitrator as to how he should thereafter conduct the arbitration'. Again, in *Carlisle Place Investments Ltd.* v *Wimpey Construction (UK) Ltd.* (1980) 15 BLR 109, Goff J, (as he then was), having held that an arbitrator may limit the amount of evidence given provided he acts fairly, stated that, generally speaking he is the master of his own procedure.

In *Bremer Vulkan Schiffbau und Maschinenfabrik* v *South India Shipping Corp. Ltd.* [1981] AC 909, Roskill, LJ, as he then was, said at p. 948 that an arbitrator or umpire who attempted to conduct an arbitration along inquisitorial lines except with the express agreement of the parties might expose himself to criticism and possible removal. Again, in *Chilton* v *Saga Holdings* [1986] 1 All ER 841 it was stated (at p. 844b) that 'both courts and arbitrators in this country operate on an adversarial system of achieving justice' and that a County Court registrar was wrong to insist that questions by a solicitor who represented one of the parties to the other (unrepresented) party should be put through him (the registrar); the mere fact that a party is unrepresented cannot deprive the other of his right to cross-examine directly.

However, the Court of Appeal in the *Bremer Vulkan* case was reversed, though on rather different grounds, Lord Diplock saying (at [1981] AC p. 985) that subject to any procedural rules which may be specified, 'the parties make the arbitrator the master of the procedure to be followed in the arbitration' adding that 'he has a complete discretion to determine how the arbitration is to be conducted from the time of his appointment to the time of his award, so long as the procedure he adopts does not offend the rules of natural justice'.

On the whole, it would not seem that the *Chilton* case has seriously impaired the flexibility of arbitral procedures, though without the parties' consent the arbitrator should be wary before embarking on an investigation himself, except in the case of 'look-sniff' arbitrations where the procedures are supported by long practice.

2. High Court procedure

Since the procedure adopted in many, perhaps most, domestic arbitrations in this country follows that of an action in the High Court with or without simplifications or other modifications, it is convenient to summarise what that procedure is. In doing so we shall assume that the parties are represented by an 'advocate' who for this purpose may be either Counsel, solicitor, accountant, civil engineer or any other person:

(a) The claimant's case is opened, together with his defence to any counter-claim, by his advocate;

(b) The claimant himself is called, assuming he is going to give evidence at all, and may be cross-examined by the other side. If he is, he may be re-examined by his advocate, but only on matters raised in the cross-examination. He may now be questioned by the tribunal after which he may be cross-examined by the opposing advocate and re-examined by his own on matters arising out of those questions (but **not** on other matters);

> **NB** The tribunal may have interposed questions earlier but strictly these should have been confined to matters on which he was then giving evidence and for clarification purposes only.

(c) The claimant's witnesses should now be called in succession, the procedure being as in (b);

(d) The respondent's advocate opens his case;

(e) The respondent is called, if he is to give evidence, the procedure following that mentioned under (b), *mutatis mutandis;*

(f) The respondent's witnesses are called in succession, with the same procedure;

(g) The respondent's advocate addresses the arbitrator;

(h) The claimant's advocate replies.

If a party is not represented the above procedure is followed so far as may be, but for obvious reasons such a party will be at some disadvantage if he wishes to give evidence himself.

3. Some arbitral modifications

In view of the considerations set out under the first sub-heading of this Chapter, there are clearly limits to the extent to which an arbitrator may modify normal court procedures if only one of the parties consents. It will, of course, be in mind that if both parties oppose a suggested manner of proceeding, or jointly require the adoption of some other method, their views normally prevail. In particular, if the parties agree that the arbitration shall be conducted in accordance with certain rules, e.g. those of some arbitral or trade body, the arbitrator has no authority to substitute other rules.

Nevertheless, it would seem that subject to any proper objection by the parties on the ground, for instance, of privilege, the arbitrator may proceed by questioning the parties and those claiming through them, and by calling for the production of all relevant documents within their power and possession. This is because he is specifically empowered to act in this rather inquisitorial fashion by S12(1) of the 1950 Act unless the arbitration agreement provides otherwise. However it would be unusual, and in cases of any complexity perhaps inappropriate, to carry this very far.

There are various ways in which an arbitrator may depart from the normal practice of High Court actions without laying himself open to any reasonable criticism that he has abandoned essential features of the adversarial system. Indeed, when such a departure is dictated by considerations of saving time and expense failure to act would be criticised.

Features of normal litigation the preservation of which is clearly *not* essential in arbitrations include:

(a) The tribunal's lack of full information before the hearing begins as to the nature of the dispute – the documents and other evidence, and the arguments on either side.

(b) Long openings by Counsel based on the assumption that the tribunal has not already studied the vital documents in the case.

(c) Extensive legal argument and citations from authorities required because adequate summaries have not been submitted to the arbitrator beforehand.

(d) Long examinations-in-chief which the exchange of proofs of evidence could have been reduced to questions on a few controversial matters.

4. Special procedural considerations

There are certain types of arbitration to which special procedural considerations apply:

(a) Where the arbitrator has been selected because he possesses some specialised knowledge or expertise, he is likely to be able to take the expert evidence much more quickly and to shorten the proceedings in other ways, in particular in regard to the pleadings and the presentation of the arguments. However, as already mentioned, if he proposes to rely on his own experience, and adopt an approach or take a point other than one which a party has already dealt with in evidence or argument, he must make this clear so that the parties have a full opportunity to put forward alternative views for his consideration – *Zermalt Holdings v Nu-Life Upholstery Repairs Ltd.* [1985] 2 EGLR 14.

(b) Many arbitrations invite the taking of short cuts whether in regard to interlocutory matters or at the hearing, or both, as, for instance, where the issues are so clearly set out in the correspondence that formal pleadings would be otiose, and/or where the arbitrator finds he can save time by stating that, subject to any further evidence or argument to the contrary, he is already satisfied on certain aspects of the matter. Again, he may be able wholly to eliminate some issues, or supposed issues, by informally exercising his skill at conciliating before or even during the hearing.

(c) Where the issues do not involve a finding on complicated facts the arbitrator may be able to dispense with preliminaries and determine the matter without more ado. Thus, where the applicability or meaning of a contractual provision is in issue but the parties agree on the monetary consequences of a decision either way, the parties merely need to advance their respective legal contentions and the arbitrator can reach a decision and make his award accordingly. A simple issue of fact, such as whether it was negligent to use a particular material, can be dealt with similarly if the effect of having used it is not in dispute.

(d) Cases may sometimes be determined without a hearing at all, the arbitrator being asked to base his decision on an inspection or on the written submissions of the parties and to award specified sums depending on whose submissions he accepts.

(e) See also section 8 below which deals with 'Documents only' arbitrations.

(f) No formality of any kind is needed where all that is required is an expert view as to whether particular goods are in accordance with sample or a given specification, as in the case of 'look-sniff' arbitrations – see p. 67.

It should also be mentioned that various modifications based on continental practice and experience have been successfully adopted in some arbitrations, and also to some extent in the Commercial Court.

5. Attendance at the hearing

Those entitled to be present at the hearing comprise the parties, those claiming under them, and their advisers, informal as well as formal. Thus, in *Haigh* v *Haigh* (1861) 31 LJ Ch. 420 (CA), an award was set aside because an arbitrator refused to allow the attendance of a party's son who helped with the accounts, and that of a shorthand writer; this was considered unfair.

The common court practice in civil cases of permitting witnesses who have not yet given evidence to be present is generally followed but may require modification where fraud is alleged or there is a conflict on factual matters as distinct from expert opinions.

For power to compel the attendance of witnesses, see at pp. 43–44, *ante*. See also para. (g) on p. 56.

6. Evidence

Arbitrators have power to examine witnesses on oath (or affirmation) but the fact that references to oaths are to be found in three successive subsections – the first three of S12 of the 1950 Act – should be attributed to indifferent drafting rather than inherent importance; in practice, many arbitrators, including members of the London Maritime Arbitrators' Association, require evidence to be given on oath in exceptional cases only.

The rules of evidence being a matter of the general law rather than that of arbitration are not discussed here but, for convenience, a copy of the Civil Evidence Act, 1968, will be found at pp. 155 *ff post*.

The arbitrator usually limits the number of expert witnesses which each side may call and directs them to exchange proofs of evidence and prepare a joint report on all relevant matters on which they agree. The experts are frequently encouraged to meet with this end in view perhaps, but not necessarily, 'without prejudice'.

Attention was drawn in Chapter 5 section 2 (p. 51) to the advantages of narrowing the technical issues at a very early stage. If it proves possible to put them into a single joint report no oral evidence from experts may be needed.

It should be noted that in England expert witnesses are expected to assist the tribunal by giving a fair view on the matters upon which their evidence is sought, not *merely* to forward the case of the party employing them. In practice, an unduly biased expert may do his side more harm than good.

7. Legal assessors

Legal assessors are occasionally appointed by a tribunal whose members are not legally qualified when it seems that a main issue will depend upon a point of law. An assessor will rarely need to be present throughout the hearing though the opportunity will probably be taken to seek his advice not only on the questions in issue but in regard to costs, taxation, interest, and perhaps the drafting of the award. In general, however, it is seldom desirable to add to the costs by making such an appointment. There is usually no difficulty in finding an arbitrator who is not only technically equipped in the relevant field but has sufficient experience to cope with most legal points which are likely to arise. Alternatives where the outcome of a specific legal issue is of vital importance are to call expert evidence from a specialist lawyer, and to apply to the Court for a ruling under S2 of the 1979 Act – see Chapter 8 – section 3(b) p. 93.

Where an assessor is appointed it will be appreciated that he is merely there to offer advice when asked; the decision must be, and be seen to be, that of the arbitrator. It is indeed desirable from more than one point of view that not only his identity but also what advice he tenders are known to the parties.

8. 'Documents-only' arbitrations

If it is practicable, and in particular if the facts are not too complex, a 'documents-only' arbitration may save much time and money. The benefits, or the reverse, depend on the circumstances, but a signal advantage in costs is certainly gained where the parties can collect the necessary documentation themselves and forward it to an agreed arbitrator without the intervention of professional advisers.

This is usually the case with the low-cost consumer arbitration schemes which, originally inspired by the Office of Fair Trading, are administered by the Chartered Institute of Arbitrators on behalf of, but completely independently from, various trade associations. At the time of writing, there are some 40 such schemes in existence and over 1,000 disputes are resolved each year. Each side makes a fixed contribution and pays his own costs for any outside advice or representation. The trade association bears any additional costs of the tribunal. The successful party will usually recover his registration fee from the unsuccessful one.

These are genuine arbitrations in that they are conducted in accordance with the Arbitration Acts and the awards are consequently enforceable in the normal way, though this is seldom necessary in the case of the trader since his compliance is likely to be obligatory under his association's rules of conduct.

Use of the scheme is at the option of the consumer who instead may elect to sue. The supplier of the service usually has no option, one regrettable exception being under the recently inaugurated solicitors' scheme which enables a solicitor against whom a complaint is made to refuse to arbitrate.

Court procedure in England is largely based on the examination and cross-examination of witnesses and certainly oral examination sometimes throws a quite new light on mere documentation. On the other hand, oral evidence given long after the event is apt to be very unreliable as well as expensive, and may add little, if anything, to available documentary evidence; besides, judging veracity by the demeanour of witnesses is a very inexact science in which some are doubtless even less skilled than others. The practice is probably at its most useful in cases of fraud but at other times may simply confuse a witness and waste time.

One compromise between documents only treatment and a full hearing is to supplement the documents with a site inspection. Another possibility is a short informal hearing at which the protagonists add oral submissions to the documentary evidence and the arbitrator can question the parties.

9. 'Look-sniff' arbitrations

'Look-sniff' or 'taste-sniff' arbitrations have always had considerable practical importance because they depend almost entirely on technical skill in a particular trade and the ordinary law courts can therefore never be a substitute.

Their procedures are governed in large measure by the customs of the trade in question, and usually there are no lawyers, speeches, witnesses or arguments. Most of the arbitrations are conducted under the auspices of a trade association which will appoint the tribunal, receive examples of the goods alleged to be sub-standard, and pass them on to the arbitrator together with the samples and/or descriptions forming the basis of the contract. As like as not the goods will be examined – although evidence or arguments will not be heard – in the absence of the parties or their representatives.

Nevertheless, the arbitrator's conclusions, subject to the usual requirements of certainty, finality and so on, will have the full status and enforceability of arbitral awards.

Almost inevitably, if there is any appeal it will be conducted under the rules of the trade association concerned.

10. 'String' arbitrations

In the case of certain commodities whose ownership frequently changes hands, the number of disputes which would otherwise require resolution are sometimes reduced by a provision for 'string' arbitrations. Thus, under Rule 5 of the GAFTA Rules, in cases where the only material difference between contracts for the sale of goods is that of price, an arbitration as to quality and/or condition may take place between the first (or a later) seller and the ultimate (or an earlier) buyer with the result binding on all parties in between. Each party claiming to be in the string must provide the arbitrator with his contract and all relevant information. Rule 12 entitles any party to appeal to the Association's Board of Appeal – except as to the condition of goods where the sale was on 'Rye terms', that is to say, on the basis of their condition on arrival – see Rule 8(1).

11. 'Pendulum' arbitrations

Pendulum arbitrations, sometimes called 'flip-flop' arbitrations, take place when the arbitrator only has power to find in favour of one side's contentions or the other's, not anything in between. It is sometimes successfully used in fair rent disputes – and also in wages arbitrations, which are outside the scope of this book. The idea is to discourage the extravagant demands which parties sometimes make with a view to obtaining a more favourable decision where they judge the arbitrator to be of the type who is apt to split differences down the middle.

In a refinement of the pendulum idea, each side makes its claim and the arbitrator, having formed his own view, decides halfway between it and the claim nearest to it.

12. Commodity arbitrations generally

Many types of dispute concerning commodities are not susceptible to 'look-sniff' treatment and a rather more formal procedure is appropriate. In the case of most commodities, or groups of commodities, an association exists representing some or all of the categories of dealer mainly involved –

growers, shippers, brokers, traders, manufacturers, and so on – perhaps on a world-wide basis. These associations very often have standard forms of contract and their own arbitration rules which they amend from time to time in the light of experience. A common feature is that disputes between members are settled internally as far as possible and without the intervention of lawyers.

For instance, under the Refined Sugar Association Rules, 1986, a member may consult the association's own lawyers and be represented at a hearing by an agent engaged in the trade, but may not appear by Counsel or solicitor without special leave and this the Association may refuse without giving reasons – Rule 8. On the other hand, the tribunal itself quite often appoints a legal assessor to give help on legal questions.

Some associations provide for arbitrations to be conducted before a panel of, say, five or even more of its members.

Others provide for the arbitration to take place in two steps – in the case of GAFTA, first, before a tribunal of three of its members, and then, if either party expresses dissatisfaction within a given number of days, before a board of appeal of five members picked from a special panel.

The rules of the association concerned may supplement the usual methods of enforcing awards by providing for the circularisation of the names of defaulters or even for their membership to be forfeit.

13. Construction industry arbitrations

Contracts in the construction industry frequently follow in whole or in part model contracts prepared by one or other of bodies such as the Joint Contracts Tribunal (JCT) – which comprises representatives from all facets of the building industry – or the Institute of Construction Engineers (ICE). The JCT contract is now in fact the RIBA contract.

Arbitral referees or 'adjudicators' are sometimes appointed but opinions are divided as to whether it is a good plan for there to be immediate references on points of detail, even important ones, during the course of a construction contract. It is argued that it greatly diminishes the authority and freedom of the architect or engineer in charge. On the other hand, it enables differences of opinion to be resolved, and avoiding action taken where necessary, at an early stage, and it makes it much easier to ascertain what exactly has been done. Besides, the existence of adjudicators who are paid on a time basis provides the parties with an incentive to settle disputes themselves. A feature of ICE contracts is that disputes are referred to the engineer for his decision before they go to arbitration.

In a substantial case, arbitrations usually follow court procedure with a few modifications designed to save time or money. Sometimes, however, where the nature of the dispute makes it appropriate, normal pleadings are discarded in whole or in part and issues are set out in the form of a 'Scott' or 'Official Referee's' Schedule. This method is very adaptable; indeed, the illustration given at p. 114 is based on a dispute which has arisen within a professional firm rather than under a building contract.

It was held in *Higgs and Hill Building Ltd.* v *University of London* (1983) 24 BLR 139 that the principles upon which leave should be given in construction cases are the same as in other cases. Indeed, arbitrations in the construction industry are usually only distinctive because of their length and complexity. The plethora of facts, drawings and other documents then becomes a special problem and every effort requires to be made to reduce the area of discovery and the size and number of agreed bundles.

Satisfactory methods of recording the proceedings can also become a problem. Even though large sums are at stake, the cost of a shorthand note in a long case is very burdensome, while tape records can be disappointing – the rustling of papers and other background noises sometimes drown the evidence while finding specific passages in the record is apt to be very time-consuming.

These difficulties can often be materially reduced by the exchange and agreement of proofs of ordinary as well as expert witnesses. Counsel can also help by submitting summaries of their addresses beforehand thereby facilitating the arbitrator's task of taking a detailed note which always slows things up.

14. High technology disputes

Certain lawyers take the view that disputes which arise from contracts for the provision of high technology are not suitable for arbitration and would prefer some form of ADR. Four main reasons are given. Breakdowns and other problems tend to be prevalent since the subject matter of the contract is often at the leading edge of the technology involved; second, the effect of such breakdowns can be catastrophic, perhaps bringing the entire production to a halt; third, the relationship between the parties is willy-nilly an enduring one and their concern is to get things going as soon as possible rather than confrontation; fourth, making good the losses caused by the defective technology may well be beyond the resources of the party who supplied it but bankrupting him may do more harm than good.

It is not the existence of these important differences from more humdrum contracts that is open to doubt, but the conclusion drawn. Certainly, neither litigation nor arbitration on normal court lines would be a solution for a string of reasons. However, mediators – 'toothless arbitrators' – and mini-trials seem little better. The most promising alternative seems to be an arbitrator with adequate technical expertise and adaptable procedures who can act as an arbitral referee immediately there is a crisis and advise the parties what steps should be taken to cure the problem and limit the damage, and who in due course can decide who, if anyone, was to blame and how the damage ought to be borne. It may be helpful in such a case, particularly where the loss is likely to exceed the contractor's resources, if the contract gives the arbitrator power to alter its terms and/or to draw a new one.

There is no unanimity as to the extent to which the original contract should detail the procedure to be adopted in the event of the serious malfunctioning of high technological equipment; much must depend on the circumstances. It is, however, clear that this aspect deserves proper attention when the contract is negotiated and should not just be a case of 'the usual arbitration clause'.

7 The Award

1. General

A valid award is the culmination of the arbitrator's work. An arbitrator will have failed in his duties if his award is not enforceable, which could be the case, for instance, if he could go back on it, or if it were tainted by illegality or impossible to perform.

The court will try to resolve an ambiguity in an award in a manner which upholds rather than destroys its validity – *Christopher Brown Ltd.* v *Genossenschaft Oesterreichischer Waldbesitzer R. GmbH* [1953] 2 All ER 1039, per Devlin, J., at p. 1,041 B. Subject to the terms of the arbitration agreement, it will be bad *in toto* if it deals with matters which were not referred, or if it fails to deal in a final manner with all disputes which were covered by the reference and actually brought before the arbitrator – with the exception, if there is concurrent litigation, of issues which will be before the Court – see *Northern Regional Health Authority* v *Derek Crouch Construction Co. Ltd.* [1984] 2 All ER 175 (CA). All material issues raised should be mentioned, even those withdrawn during the hearing, in order to prevent their being brought up in other proceedings.

However, if there are a number of issues, an award cannot be impugned simply because each has not been expressly dealt with provided it is reasonable to infer that all were taken into account.

The arbitrator must reach a decision on the actual questions put; it is not sufficient merely to devise a fair way of dealing with the problems which have arisen.

He is not obliged to accede to the insistence of a party that particular documents should be annexed to the award – *Mafracht* v *Parnes Shipping Co. SA The Apollonius* [1986] 2 Ll. Rep. 405.

While an award may direct one party to convey property to another it does not itself effect the transfer unless the arbitration takes place under a statute which so provides.

The arbitrator has the same power as the High Court to order specific performance of any contract, other than one relating to land or an interest in land, unless the arbitration agreement takes it away – S15 of the 1950 Act.

2. Finality of an award

Every arbitration agreement, unless the contrary intention is expressed, is deemed to provide that the award is to be final and binding on the parties and on those claiming under them – S16 of the 1950 Act.

Unless and until its validity is successfully attacked, it is conclusive as between the parties of the facts found by it, for example, as to title of land, but does not constitute evidence against a stranger or in criminal matters.

When a final interim award is made as in *Chiswell Shipping Ltd. v State Bank of India* (No. 2) [1987] 1 Ll. Rep. 157, the arbitrator is *functus officio* as regards the issues with which it deals but he retains jurisdiction in regard to other issues.

As mentioned on p. 80 even after making a final award he can still be called upon to provide for the costs if he has not done so – see S18(4) of the 1950 Act.

In addition, unless the arbitration agreement provides otherwise, he retains power under S17 of the 1950 Act to correct a clerical mistake or error in the award which arises from an accidental slip or omission. Such an error must be one affecting merely the expression of the arbitrator's thought, not an error in the thought process itself, since he cannot reconsider his award – see *Mutual Shipping Corp. of New York* v *Bayshore Shipping Co. of Monrovia* [1985] 1 All ER 520 (CA) and *Food Corporation of India* v *Marastro Cia Naviera SA. The Trade Fortitude* [1986] 3 All ER 500 (CA). In the former case it was pointed out that the court has power to remit an award under S22 of the 1950 Act which it may exercise in exceptional circumstances; if an arbitrator is in any doubt as to whether S17 applies he should apply to the court himself.

The court may construe an award, and may set it aside either if it was improperly procured or the arbitrator (or umpire) misconducted himself. Any person applying to have an award set aside may be ordered to bring into court, or otherwise to secure, any amount payable under it – S23(3) of the 1950 Act.

The court may remit the matters referred for reconsideration by the arbitrator, for example, where he requests it because of a mistake, as mentioned in the *Mutual Shipping* case, *op.cit.*, or because new evidence comes to light, but may no longer remit the award on the grounds of an error of law or fact on its face.

Finally, the award may be appealed subject to the restrictions mentioned in Chapter 8.

3. Interim awards

Subject to the provisions of the arbitration agreement, interim awards may be made. 'Award' includes 'interim award' wherever it is mentioned in Ss1–34 of the 1950 Act – S14, *ibid*.

Interim awards are often used when the arbitrator has determined who is liable and the parties indicate they expect to be able to agree upon the damages. This will cut down the costs substantially if all goes according to plan while, if they fail to agree, or either of them wishes the result to be incorporated in an award for enforcement purposes, the matter can be referred back to the arbitrator who will have retained his jurisdiction to make a final award.

Indeed, it may save time and money to make an interim award on questions of liability since to the extent that claims fail questions of quantum may become irrelevant. Besides, the elimination of certain issues may facilitate the settlement of others. However, this course is by no means always appropriate especially where evidence as to quantum is closely linked to that of liability when if a claim is sustained witnesses may have to be re-called.

An award which disposes finally of all issues of liability but reserves issues of quantum to a further award may be referred to as an interim final award – see *Trave Schiffahrts GmbH* v *Ninemia* [1986] 2 All ER 244, (CA).

If there is no substantial defence to part of a claim and the claimant is suffering serious prejudice pending an award, the arbitrator may agree to consider this aspect initially and, having reached a decision on liability and quantum, make an interim award. In this neat way the benefits of accelerated court procedures can effectively be enjoyed without foregoing the customary advantages of arbitration.

4. Reasoned awards

The parties cannot compel the arbitrator to give reasons except through an application to the Court pursuant to S1(5) of the 1979 Act—where it applies – but nowadays reasoned awards are given in most cases of substance. After all, the parties will have worked long and thought anxiously about the matter and in deference to them, and their advocates, if any, they should be enlightened as to the processes of thought which have led to their success or failure as the case may be. These remarks would not apply to 'quality' arbitrations where no point of law arises, for example, as to whether goods accord with a sample or description supplied, nor where the amount involved is so small that saving cost is of paramount importance.

The policy of the 1979 Act was to encourage reasoned awards. In *Warde* v *Feedex International Inc.* [1984] 1 Ll. Rep. 310, Staughton, J, said they should be given if a party asks for them, except in very exceptional cases. If both ask that none should be given there should be no reasoned award but reasons should still be provided in a separate document on request. If one party asks for there to be no reasoned award and the other says nothing, no reasons should normally be given but the latter may not be aware of his rights and the arbitrator should consider whether it would be right to enquire further. Finally, if neither party says anything but both are represented by experienced advocates, the arbitrator would be justified in assuming that the parties wanted an award that would be final, that is to say, one without reasons and therefore unappealable.

Other important considerations to be taken into account if they have any bearing on the matter are the terms of the arbitration agreement and any subsequent agreement between the parties, the provisions of any rules applicable to the arbitration in question, and whether the giving or withholding of reasons would cause hardship to one of the parties.

The arbitrator should always have in mind that under the 1979 Act the purpose of giving reasons is not to inform the parties but to enable the court to judge whether a question of law arises which merits granting leave to appeal. Accordingly, he should explain briefly the contentions he has rejected, and why, but need say very little if anything about why he reached a given decision on disputed facts.

It also follows that there may be exceptional cases where the arbitrator may think it appropriate that the additional costs of producing a reasoned award should fall on the party asking for it.

If given, the reasons should not be unnecessarily technical or complex but rather should simply state in plain language why the arbitrator reached the conclusions he did – see *Bremer Handels GmbH* v *Westzucher GmbH* [1981] 2 Ll. Rep. 130 (CA). Lord Roskill's gentle rebuke in regard to the 96 pages of reasons in *Antaios Cia* v *Salen Rederierna* [1984] 3 All ER 229, should also be borne in mind, namely, that in general businessmen are interested in the decision rather than the philosophy which underlies it – see at p. 239.

Since the 1979 Act came into force, even the most optimistic party is likely to ask for a reasoned award in case the finding should unexpectedly go against him because if neither party does and no reasons are in fact given it will usually be too late to apply to the court – see S1(6) and p. 94.

With some arbitrators, preparing a reasoned award can be a lengthy and therefore expensive affair and it could save costs in a complicated case if the parties were able to say that reasons would only be required on certain specified aspects.

Lord Justice Bingham's comprehensive discourse on reasoned awards delivered on 5 October 1987 at Lincoln's Inn under the title 'Judicial Techniques' is well worth study. It is published by the Chartered Institute of Arbitrators.

At one time some arbitrators were reluctant to give a reasoned award because of the opportunity it gave the losing party to pick over the finding and perhaps gather enough material to support an appeal. This risk is much less serious since the abolition by the 1979 Act of both the case stated procedure and the court's ability to set aside or remit an award on the ground of an error of fact or law on its face; and it does not exist where there is an exclusion agreement. However, in spite of these considerations and the limitations set on appeals by S1 of the 1979 Act and the *Nema* guidelines, reasons in some London maritime arbitrations are still supplied to the parties separately from the award itself, in confidence, i.e., on terms that they are not to be referred to in subsequent litigation between the parties.

It seems that the Court may insist on looking at confidential reasons. See *Mutual Shipping Corporation* v *Bayshore Shipping Co. of Monrovia, The Montana* [1985] 1 Ll. Rep. 189 (CA) per Donaldson M. R., at p. 192, though he said of such reasons in *Trave Schiffahrts* v *Ninemia* [1986] 2 All ER 244, at p. 246: 'Naturally we know nothing of their content'.

5. Form

A submission may require an award to be simply declaratory as to the rights of the parties. Otherwise, the arbitrator should confine his directions to the performance of definite acts such as paying a sum of money or handing over specific property so as to facilitate enforcement should that prove necessary. Instructions to enter into contracts the terms of which are left in the air and, above all, the inclusion of pious hopes, should be firmly eschewed.

It is desirable to spell out the formalities unless the chances of a challenge to the award are remote. It is true that in *Smith* v *Hartley* [1851] 10 CB 800, it was said that the arbitrator will be presumed to have had jurisdiction and that anyone contending the contrary must prove it, but in *Brown (Christopher) Ltd.* v *Genossenschaft Oesterreichischer Waldbesitzer* [1954] 1 QB 8, Devlin J, expressly denied that there is any presumption of validity in the case of arbitral proceedings.

Legal assistance may be sought in drafting the award but not in making it. However, a purely ministerial act as distinct from a judicial decision may be left to a third party. Thus, in a question of valuation an arbitrator may determine that the rate is so much per acre directing that the number of acres is to be determined by measurement, an application of the old Latin tag: '*Id certum est quod certum reddi potest*'.

A convenient order for an award in a complex matter is as follows:
(a) A background note of the relationship between the parties out of which the dispute has arisen;
(b) Reference to the arbitration agreement and the appointment of the arbitrator;
(c) The main steps in the arbitration, with dates;
(d) A summary of the issues;
(e) Some reference to the vital witnesses and evidence;
(f) The arbitrator's findings of fact;
(g) The arbitrator's conclusion on any issues of law and their application to the facts found;
(h) The arbitrator's award including his decisions on interest and the costs of the reference.

An illustration will be found on p. 116 *ff*.

For obvious reasons, an award should normally be in writing, dated and signed – indeed, the signature should preferably be witnessed. Theoretically, however, no signature is necessary.

Where an umpire has been appointed but has not entered upon the arbitration – for instance, because the arbitrators have not disagreed – the arbitrators alone should sign the award but the addition of the umpire's signature would not of itself invalidate it. Similarly, if the umpire has entered upon the reference it will be his award but it will not be rendered void merely because the arbitrators add their signatures.

If there is more than one arbitrator all must participate in the actual decision but a defect on this score cannot be attacked by a party who has accepted a benefit under part of the award – *European Grain and Shipping Ltd.* v *Johnston* [1982] 3 All ER 989.

It is no longer considered necessary for all to sign at the same time and place – *ibid.*, see at p. 992; the practical solution if the arbitrators are in different places is to pass round drafts and agree the contents, and then circulate an original and copies for signature. NB S9 of the 1950 Act, as substituted by S6(2) of the 1979 Act, provides for a majority decision where there are three arbitrators.

Where arbitrators act as advocates after an umpire has been appointed – see p. 39 – the award should distinguish their fees for performing judicial functions from those referable to their functions as advocates – per Megaw, J, in *Government of Ceylon* v *Chandris* [1963] 2 All ER I.

6. Currency

Normally an award should be expressed in sterling but in these days of fluctuating exchange rates it is recognized that this is not obligatory even in a domestic arbitration and the arbitrator should use some other currency if this is necessary to do justice between the parties – *Services Europe Atlantique Sud* v *Stockholms Rederiaktiebolag Svea, The Folias* [1979] AC 685. Where there is a choice of possible currencies, the general rule is that an award should be made in the proper currency of the contract, i.e. the one with which payments under it have the closest and most real connection. If this is in doubt it should be made in whichever currency seems to produce the most just result. In *The Folias* case, French charterers had to use francs to buy cruzeiros to recompense Brazilian importers for some damaged onions and it was held that the award should be in French francs notwithstanding that the hire charge and various other payments were to be made in US dollars. See also the illustration in Appendix 1(d) paras 12(iii) and 16(c) at pp. 118 and 119.

7. Timing

In practice, the arbitrator will save a great deal of time if he begins to draft his award immediately after the hearing while the details are still fresh in his mind, but there is only one actual time limit, namely, where the court remits an award for re-consideration. Then the arbitrator (or umpire) must take the necessary action within three months unless the order otherwise provides or the time is enlarged; in other cases, subject to anything in the arbitration agreement, an award may be made 'at any time' – see Ss 22(2) and 13(1) of the 1950 Act respectively. See, also, S1(2) of the 1979 Act.

The court may extend any period which is laid down even after it has expired and even though it was laid down in another statute – S13(2) *ibid*; and see *Knowles & Sons Ltd. v Bolton Corp.* [1900] 2 QB 253, concerning an arbitration under the Public Health Act 1875. The parties may always extend the time by agreement and will be deemed to have done so if they acquiesce in the continuance of proceedings after expiry of the time within which the award should have been made.

There is, however, one general limitation on what could otherwise be excessive licence; an arbitrator or umpire must use 'all reasonable despatch' throughout, failing which he may be removed by the High Court and lose his remuneration – S13(3) of the 1950 Act.

The making of an application to remove an arbitrator does not prevent him making a valid award – *Oakland Metal Co. Ltd. v Benaim (D) & Co.* [1953] 2 QB 261.

It is sometimes important to distinguish between the date of making the award, i.e., its signature, which, by an unfortunate piece of nomenclature is sometimes referred to as its 'publication', and the date of its actual publication to the parties. It is within 21 days of the latter date that an application to the court to remit an award must be made under o.73 r.5.

8. Costs

(a) Generally

The 'costs of the award' comprise the arbitrator's fee, together with his expenses such as travelling, the cost of any legal advice he may take, and of hiring a room for the hearing (if paid for by him), plus VAT if applicable. Depending on the context, the 'costs of the reference' may mean either the costs which one of the parties will have to pay if costs are given against him, that is to say, those properly incurred by the parties in the course of the arbitration, alternatively, such costs excluding the costs of the award.

Parties contemplating arbitration quite often reach an arrangement about the costs beforehand the most common being that each side will bear its own together with one-half of 'the costs of the award'. However, any such arrangement having the effect of rendering a party liable to pay the whole or part of the costs in any event is void if it is made before a dispute arises, the arbitration agreement then taking effect as if no such provision were in it – S18(3) of the 1950 Act.

A valid arrangement, for example, one made when the parties are already in dispute, will bind the arbitrator. Otherwise, unless the arbitration agreement otherwise provides, the costs are within his discretion by virtue of S18(1) *ibid.*, and it is his duty to exercise it. See paragraph (c) below.

If the award makes no provision for costs, any party may apply to the arbitrator for a direction within 14 days of its publication and he must amend the award accordingly after hearing any party wishing to be heard. The court has power to extend the 14 days – S18(4) of the 1950 Act. It should be noted that a statement in the award that the arbitrator makes no order as to costs *is* a provision for the costs; it is simply a way of saying that each party must pay his own.

In a complicated case, he may well find it convenient to make an interim award, fix an appointment to hear argument on the question of costs, and incorporate his decision on the latter in a final award. This not only enables those concerned to concentrate on a single, sometimes difficult, issue but has the advantage that once the question of liability is out of the way a party can disclose any offer he has made during negotiations without fear that it may prejudice his case.

It is worthy of note that Parliament has placed solicitors in a privileged position by giving the court power to charge property recovered or preserved in the arbitration with payment of their costs (though nobody else's) – S18(5) *ibid.*

It remains to point out that any order made under Part 1 of the 1950 Act may be on such terms as to costs as is thought just – see S28, as amended by the 1975 Act, S8(2)(b).

(b) Taxation

Having directed by whom the costs are to be paid the arbitrator, again subject to the arbitration agreement, may tax[1] or settle their amount.

Section 18(1) of the 1950 Act provided that the costs awarded might be those 'as between solicitor and client' subsequently called the 'common fund basis', which was more onerous than the commoner 'party and party' basis and usually only awarded when there was some default by the party liable.

In 1986, however, amendments to Order 62 of the RSC swept these types of cost away in favour of 'costs on a standard basis' or 'costs on an indemnity basis'. The former comprise a reasonable amount for all costs reasonably incurred, any doubts to be resolved in favour of the person liable to pay the costs. Costs on an indemnity basis, on the other hand, comprise a reasonable amount for all costs reasonably incurred with any doubts resolved in favour of the person entitled to receive them. Which basis to use is a matter for the arbitrator's discretion provided always that he acts judicially, as to which, see p. 82 *post*.

Section 18(2) goes on to provide that any costs which an award directs to be paid 'unless the award otherwise directs' are taxable in the High Court. An award would direct 'otherwise' if it undertook taxation itself or provided that the costs were to be taxed in the County Court – see *H. G. Perkins Ltd.* v *Brent-Shaw* [1973] 1 WLR 975.

Among the costs the arbitrator may tax, subject to the terms of the arbitration agreement, is his own remuneration, which should be shown separately from other expenses.

An umpire is in a similar position except that the costs he may tax will include those of the arbitrators. He does not do this by merely incorporating what they ask in his award but must exercise an independent mind taking into account the legitimate interests both of the arbitrators and of the party on whom the costs will fall. A similar approach is necessary where an arbitrator (or umpire) taxes his own fees – *Government of Ceylon* v *Chandris* [1963] 2 All ER 1. Failure in this regard amounts to technical misconduct.

If the arbitrator does not feel qualified to tax the costs the task may be delegated to the High Court, or indeed the County Court, unless the arbitration agreement requires him to settle them. The former would be effected by inserting in the award: 'Costs if not agreed to be taxed on the [standard] basis in the High Court' in which event taxation would presumably be on the High Court scale even though the amount would have been within the County Court's powers.

Taxation by the High Court involves proceedings before a taxing master which are costly, cumbersome and involve considerable delay. It is even worse where the decision is appealed, though this does not often happen. It may therefore well be that the Costs Arbitration Service which the Chartered Institute of Arbitrators set up in 1986 will prove a considerable boon to all concerned, especially in complicated cases. It may consist merely of advice from an experienced assessor which it would seem from *Rowcliffe* v *Devon and Somerset Railway Co.* (1873) 21 WR 433, an arbitrator is entitled to take, or be a mini-arbitration on its own, a course to which the parties would have to agree. However, in a straightforward case taxation is not quite so formidable a task as may appear at first sight. The party in whose favour an order for costs has been made will produce an itemised bill of costs and the other side an itemised criticism. Each party will no doubt support his viewpoint with argument. Clearly an experienced arbitrator having heard the case and read most of the documents ought to be in a better position to judge between competing claims than a taxing master with no previous knowledge of the case.

(c) The arbitrator's discretion

The arbitrator must exercise his discretion in regard to costs 'judicially' and not capriciously; provided he does, the court will not substitute its own discretion even though it would have decided otherwise itself.

It follows that he should apply the normal rule that costs follow the event i.e., that the loser pays, unless there is good reason to the contrary.

Where there *is* good reason, it is highly desirable to make it clear in the award since this may well avoid an aggrieved party incurring costs and trouble attempting to get the award set aside – see the second of Mocatta, J's propositions in *The Erich Schroeder* [1974] 1 Ll. R. 192 at p. 193.

In *Warinco A. G.* v *Andre et Cie SA* [1979] 2 Ll. R. 298 Donaldson, J., (as he then was) expressed the view that where there is an order which on the face of it is unusual there is a presumption, rebuttable by evidence, that the arbitrator was wrong.

In general, the question is not who has been proportionately the more successful but whether the claimant was justified in bringing proceedings provided his claim was not so inflated as to deter the other side from seeking a settlement or to increase the costs in some other way – see *Tramountana Armadora SA* v *Atlantic Shipping SA* [1978] 1 Ll. R. 391.

Sometimes, however, the question is the rather different one of determining an appropriate figure, for example, a rent increase, where the parties cannot agree. The solution then may be to award to a party who was only marginally wrong his costs with a fractional diminution or, where the arbitrator's decision is somewhere in the middle, to make no order.

A typical reason for depriving a successful party of at least part of his costs is that he has prolonged the case or otherwise materially increased the costs by an excessive production of documents, or by introducing arguments or evidence which were irrelevant to the contentions on which he has succeeded; and, *à fortiori* where he fails on a substantial aspect of the case, for example, losing on the defence but winning on a counterclaim. In the latter case the claimant is sometimes awarded the costs of the claim and the respondent those of the counterclaim, but in view of the difficulties of apportionment where the issues are intertwined it is usually better to give each party his costs or a proportion of them.

The award of the full costs to a successful party notwithstanding his failure on certain aspects of the case does not necessarily justify remission to the arbitrator – *Higgs and Hill Building Ltd.* v *University of London* (1983) 24 BLR 139; there should be no interference with his discretion except in a plain case.

There have been one or two exceptional cases where the courts have upheld awards depriving a successful party of the whole of his costs. Thus, in *Heaven & Kesterton Ltd.* v *Sven Widaeus A/B* [1958] 1 WLR 248, an arbitration agreement stated that a decision on the costs should take into account the correspondence and the parties' efforts to reach a fair settlement. The claimant won but only to the extent of one tenth of the amount claimed; costs were awarded against him since a more reasonable claim might have avoided the arbitration altogether. See, also, *Gray* v *Lord Ashburton* [1917] AC 26, where a landlord who recovered £71 out of a claim for £744 dilapidations was directed to pay the whole of the costs relating to quantum together with his own costs on the question of liability.

Where there is no clear 'winner', a convenient way of avoiding taxation which can sometimes be adapted to suit other circumstances is to direct each side to pay its own costs plus one-half of the costs of the award.

(d) Offers prior to the arbitrator's determination

In *Stotesbury* v *Turner* [1943] KB 370, it was held that the claimant's refusal during negotiations to settle on the basis of a 'without prejudice' offer in excess of the amount he was finally awarded did not justify depriving him of his costs.

However, in the case of an open offer, whether made during negotiations or the reference, the normal rule applies, namely, that if the claimant is awarded less than the amount offered he is entitled to his costs up to the date of the offer but must pay all those incurred subsequently. The concept was refined in the *Tramountana* case, *op.cit*. The arbitrator should ask himself whether the claimant has achieved more by refusing the offer assuming interest is taken into account by adding to the amount offered notional interest for the period from the date of the offer to that of the award. Bernstein[2] suggests that this period should begin at the time it would have been reasonable to accept the offer rather than when it was made.

It was pointed out in the *Tramountana* case that the difficulty that the arbitrator, unlike the judge, will usually know of the existence of a 'sealed offer' could be avoided by his calling on the respondent to hand in a sealed envelope containing *either* an offer to settle *or* a denial that such an offer has been made. It is not known how much this interesting suggestion by the present Master of the Rolls has been used but it would not seem to work unless the arbitrator makes such a direction without being prompted.

(e) Calderbank offers

A Calderbank offer is an offer made without prejudice as to all matters save costs or, more fully, a without prejudice offer subject, if it is rejected, to the right to bring it to the attention of the arbitrator on the issue of costs. Consequently, it may only be disclosed after the arbitrator has issued an award on all matters other than costs. *Prima facie*, the offeror, who may be claimant or defendant, will be entitled to all costs incurred after the offer provided the other party has gained nothing more, taking interest into account, by continuing with the arbitration instead of accepting the offer.

To avoid frustration of the whole device the party making the offer should ask the arbitrator for an interim award dealing with the merits leaving the parties to agree costs when they have read it. If they fail to agree the arbitrator will be asked to make a final award.

In *Cutts* v *Head* [1984] Ch 290, the Court of Appeal stated that a Calderbank offer was appropriate in all cases other than a simple money claim where money could be paid into court.

(f) VAT upon costs

Where the arbitrator is registered for VAT purposes VAT is chargeable on the costs except in so far as the person by whom they are payable is an overseas party; to the latter extent they are zero-rated. A party liable under the award and also to VAT will be charged the VAT attributable to his share. A party taking up an award should be given by the arbitrator receipted tax invoices for the proportion of costs (plus VAT) which each UK party is due to pay. In so far as he is not himself liable he will hand over the relevant invoice against payment by the party who is.

Where the arbitrator is not registered he will not, of course, add VAT to his costs but may pass on as part of his charges to the party or parties liable that party's proportion of any amounts upon which VAT has been borne.

9. Interest

Unless an arbitration agreement provides to the contrary, it is deemed to give the arbitrator (or umpire) power, which he should generally exercise, to award simple interest for any period up to, but not after, the award on any sum included in it plus any sum included in the reference but paid before the award – S19(A) of the 1950 Act added by S15(6) of the Administration of Justice Act, 1982. The latter rectified a very unsatisfactory situation which arose previously, namely, that a person expecting to lose could pay a sum equal to his anticipated liability just before issue of the award thereby escaping liability for interest on it – see *President of India* v *La Pintada Cia Navegacion SA* [1985] AC 104 in which Lord Brandon set out the chequered history of the legislation and court decisions on the subject of interest.

General damages for late payment of a debt are not recoverable – *London, Dover and Chatham Railway* v *South Eastern Railway* [1893] AC 429.

Special damage by reason of late payment may be claimed provided the damage arises from circumstances in the contemplation of the parties when the agreement was entered into; interest on such a sum can be awarded. See *Wadsworth* v *Lydall* [1981] 1 WLR 598 (CA), where the defendant failed to pay an agreed sum which he knew or should have known the plaintiff needed for a purchase. It was held that the additional cost of alternative finance was recoverable.

A claim for demurrage is of the nature of damages and not debt, and there is no such thing as an action for the late payment of damages. An exchange loss by reason of late payment is irrecoverable – *President of India* v *Lips Maritime Corp.* [1987] 3 All ER 110 (HL).

If there is judgment for an amount of interest it becomes a debt and interest can be awarded on it since, though interest upon interest, it is of the nature of simple not compound interest – *Coastal States Trading (UK) Ltd.* v *Mebro Mineraloelhandelsgesellschaft mbH* [1986] 1 Ll. Rep. 464.

Some forms of contract specifically provide for the payment of interest – see for instance, clause 60(6) of the 1973 ICE form.

The rate awarded should bear some relation to the relevant circumstances. Thus, it might be based on the commercial rates prevailing during the period or on an actual overdraft rate paid by the claimant.

Interest for periods *after* the award is dealt with by S20 of the 1950 Act. It provides that an amount payable under an award carries interest from the date of the award at the same rate as a judgment debt unless the award otherwise directs. It has been held by the House of Lords that the last phrase does not entitle the arbitrator to select some other rate – see *Timber Shipping Co. SA v London and Overseas Freighters Ltd.* [1972] AC1.

At the time of writing (October 1987) the rate for judgment debts is 15% by virtue of the Judgment Debts (Rate of Interest) Order, 1985. Once determined in relation to a particular award the rate does not vary with amendments to the order – *Rocco Giuseppe et Figli v Tradax Export SA* [1983] 3 All ER 598.

10. The arbitrator's lien

Whether the fee is agreed or not it is usual for the arbitrator (or umpire) to encourage, if not ensure, its payment by exercising his right of lien over the award. This right extends to valuations, etc., which he has obtained for his own guidance but not documents, etc., which have been put in evidence.

When the award is ready he will notify the parties, stating the amount of his charges; either party may then take it up. If and to the extent that the party who does so is not liable for the costs, he may claim reimbursement from the party who is.[3] However, if the latter can show that the fee was excessive he need only pay a reasonable amount and the former must reclaim the balance from the arbitrator as money had and received. Another hazard of paying a fee demanded by the arbitrator is that the fact of payment does not prevent its being taxed when claimed from the other side if the amount is not in terms taxed by the award.

Where an arbitrator or umpire refuses to deliver his award until payment of remuneration which is considered excessive, the best solution, when available, is for a party to apply to the High Court to order its delivery under S19(1) of the 1950 Act upon the applicant paying the fee demanded into court when it will be taxed in the presence of the arbitrator (or umpire) and any excess returned to the applicant.

However, this procedure cannot be used where another party has taken up the award nor where the amount involved was fixed by agreement in writing – not for instance at so much per hour – see S19(2). The only means then of challenging the amount is by application to the court either to remit the award for re-consideration of such part of it as relates to the fee or to set it aside in whole or in part on the ground of the arbitrator's (or umpire's) misconduct – see Ss 22(1) and 23(2) respectively. The latter course is possible because where there is no agreement as to the fee the implied contract that he is to be paid a reasonable amount works both ways.

To establish misconduct a strong case would have to be made out – dishonesty, virtually, in order to get the whole award set aside – and the court may order the applicant to bring into court or otherwise secure any money payable under the award – see S23(3).

Finally, if neither party takes the award up the arbitrator will be left with his contractual right to payment for his services, as to which, see p. 45.

11. Enforcement

(a) Under S26 of the Arbitration Act 1950 (as amended)

Where it is available the most convenient way to enforce an award is to use the summary procedure provided by S26(1) of the 1950 Act, as amended by S17(2) of the Administration of Justice Act 1977, whereby, with leave of the High Court, judgment may be entered in the same terms as the award and enforced in the same manner as a judgment. The practice is set out in RSC 0.73 r.10 – see *post.* at p. 177. If it is so entered, a bankruptcy notice can be founded on it. In addition, the award merges in the judgment and can no longer be enforced separately. NB A *foreign* judgment is not treated as putting an end to an award by merger.

Action by the court under this Section is discretionary in the case of a domestic award but may be mandatory where it is foreign – S36(1) of the 1950 Act.

If the award is within the County Court limit (currently £5,000) and the County Court so orders, it is recoverable by execution, or otherwise, as though payable under an order of the County Court – subsection (2) of S26.

Applications under subsections (1) and (2) are mutually exclusive – S26(3).

S26 cannot be used:
  (i) if there is no 'arbitration agreement' as defined, for example, the agreement is oral;
  (ii) if there is a substantial question as to whether the award is valid;
  (iii) where an English Court would not make an order to the same effect;
  (iv) if the award is a statutory one which falls to be enforced in the manner prescribed by the relevant Act;
  (v) where the award is only declaratory of the amount payable and there has been no adjudication as to who is liable to pay it.

See also the defences mentioned under (b) below.

(b) Enforcement by action

An award is inherently enforceable as a new contract based on an implied agreement between the parties that it shall determine their rights *inter se*; their original rights then disappear – a case of accord and satisfaction 'by substituted agreement'. It follows that in general an award is not open to objections which could have been raised in an action on the original contract, e.g., that the latter was not sufficiently stamped – *Norske Atlas Insurance Co. Ltd.* v *London General Insurance Co. Ltd.* (1927) 43 TLR 541.

In order to enforce an award by action, it is necessary to prove affirmatively that it is valid, that is to say, that the contract containing the arbitration agreement was made, that a dispute arose which fell within its terms or was otherwise duly submitted to arbitration, that the arbitrator was validly appointed, that he made the award pleaded, and that such award has not been performed. Naturally some if not all of these points may be admitted in any defence to the action. **NB** The presumption that everything has been duly carried out – *omnia praesumunter rite esse acta* – does not apply to arbitrations – *Christopher Brown Ltd.* v *Genossenschaft Oesterreichischer Waldbesitzer R. GmbH* [1954] 1 QB 8.

Defences include:
- (i) Lapse of time. See Chapter 1 under the heading 'Limitation Acts', on pp. 7–8.
- (ii) Lack of jurisdiction; for example, that the arbitrator did not have some qualification required by the arbitration agreement.
- (iii) Revocation of the arbitrator's authority before the award was made.
- (iv) Attachment by a third party of the sum awarded.
- (v) Termination of the relationship between the parties – *Bellshill & Mossend Co-operative Society Ltd.* v *Dalziel Co-operative Society Ltd* (1960) AC 832, where an award defining the areas in which members of a union of co-operatives could trade ceased to be effective against a member when he resigned his membership.
- (vi) Performance or other satisfaction of the award. Misconduct by the arbitrator is not a defence; the proper remedy where misconduct is alleged is to move to set the award aside.

(c) Suing for a declaration or for specific performance

It is technically possible to sue for a declaration that an award is binding.

It is also possible to bring proceedings to enforce a right which is established by an award – see *Bloemen Property Ltd.* v *Gold Coast City Council* [1973] AC 115 at p. 126 (PC) but an award of damages supersedes the original cause of action.

Finally, the Court may order specific performance of the contract to abide by the award, though this is more expensive and usually no more efficacious than proceeding under S26. The Court will also order specific performance in accordance with normal principles where damages would not be an adequate remedy, but not where there has been unreasonanble delay, technically 'laches', in applying for it – *Eads* v *Williams* (1854) 24 LJCh. 531. Specific performance was not ordered where there had been no part performance and the award was not enforceable because the conditions of the Statute of Frauds were not satisfied – *Walters* v *Morgan* (1792) 2 Cox Ch. Cas. 369. Performance by one party of his obligations under the award would be part performance for this purpose.

The arbitrator himself has inherent power – unless the agreement otherwise provides – to order specific performance of any contract other than one relating to land or an interest in land – S15 of the 1950 Act.

Notes

1. In this context, 'tax' simply means determine the proper sum which should be allowed for expenses; it has nothing to do with the Inland Revenue. A taxing master will, for instance, cut down a solicitor's bill if he thinks he has overcharged, or disallow a silk's fee if he considers that only junior counsel should have been briefed.
2. *Handbook of Arbitration Practice*, p. 138.
3. It is wise for the award to state in terms that if it is taken up by the party to whom costs have been given the other shall re-imburse him.

8 Appeals to, and control by, the Courts

1. The Commercial Court

The control of arbitration is exercised almost exclusively by the Commerical Court, taking up some 15 per cent of its time. This court, to which at the time of writing five judges are assigned, sometimes adopts procedures similar to those of the civil law such as reducing the oral content of proceedings by encouraging proofs to be exchanged, reading the pleadings beforehand[1], and accepting written submissions on the law with a view to shortening advocates' addresses. It has nevertheless not always attained its declared object of making proceedings before it both speedier and cheaper than in other courts.

A procedural point to note is that judges of the Commercial Court deal with interlocutory aspects of their cases themselves instead of leaving them to Masters.

The Court's functions in relation to arbitration have been summarised by one of its judges as follows:

(a) It oversees the initiation of the process so far as necessary, for example, by granting a stay of proceedings brought by one of the parties where there is a valid arbitration agreement, and by appointing an arbitrator or umpire where consensual arrangements for doing so do not exist or have broken down – see Chapter 3, 'Applications for a stay', p. 30, and Chapter 4, 'Appointment by the Court', p. 41.

(b) It makes orders in support of the arbitral process, e.g. for security of costs, and preserving or otherwise dealing with property which is the subject matter of the reference – see S12(6) of the 1950 Act. The general effect is to give the arbitrator teeth in carrying out his functions though he will in fact possess most of the powers in question if the arbitration is conducted in accordance with one of the standard sets of rules – see Chapter 4, 'Powers exercisable in an arbitration, p. 43, and Chapter 5, p. 51. The application will normally be made by one of the parties, but the Court seldom makes an order if it is apparent that the arbitrator does not desire the power in question.

(c) The Court safeguards the integrity of the arbitral process. Thus, where an arbitrator has failed to disclose an interest of such a nature as to be likely to prejudice his impartiality, or where he has evinced a clear bias for or against a party, the Court will remove him. It will also do so if he bases a finding upon grounds which, whether founded on his own experience or not, have not been mentioned at the hearing.

(d) Subject to the provisions of S1 of the 1979 Act, the Court will upset on appeal an award which it considers was based on a wrong view of the law, but only with reluctance and if a strong case is made out – see sub-paragraph 3 (g) *post*, on p. 96. The parties may however agree in certain circumstances to exclude any right of appeal – as to which see p. 99.

(e) A judge of the Commercial Court may, if he thinks fit and is released by the Lord Chief Justice, act as arbitrator or umpire in a commercial arbitration. See Chapter 4, pp. 42–3.

(f) Finally, the court will make an order giving an award the same effect for enforcement purposes as a judgment of the court.

The court's general approach to arbitrations is that they try to uphold awards, reading them in a reasonable and commercial way rather than endeavouring to pick holes – per Bingham, J., in *Zermalt Holdings SA v Nu-Life Upholstery Repairs Ltd.* [1985] 2 EGLR 14.

2. Appeals under the arbitration agreement

Before dealing with appeals to the courts, it should be pointed out that arbitration agreements sometimes impose their own system of appeal, notably where they apply the rules of a trade association which desires to deal with all disputes internally. An award in such a case may amount to little more than a survey of the facts together with the preliminary conclusions of the arbitrator(s), any party then being free to take the matter before the body which has reserved to itself the real power. This may comprise perhaps the officers or executive committee of the association or club concerned, or a sub-committee thereof, and the rules will usually pronounce that its decision is final.

Illustrations are the Boards of Appeal instituted by the arbitration rules of GAFTA and the Cocoa Association of London Ltd.

3. Appeals under the Arbitration Act, 1979

(a) A historical note

The Arbitration Act 1979, which is drafted with considerable tortuousness, applies to arbitrations commenced on or after 1 August, 1979. It greatly modified the procedure by which a party can obtain relief in respect of a mistake of law. In particular, it repealed S21 of the 1950 Act under which the arbitrator (or umpire) was entitled, or could be ordered by the court, to state any question of law arising during the reference, or from his award, in the form of a Special Case for the opinion of the High Court; it also abolished the court's power to remit or set aside an award on the ground that an error of fact or law appeared on its face – S1. Also, with some important exceptions, it enabled the parties to prohibit or restrict rights of appeal by means of 'exclusion agreements' – see p. 99.

The Case Stated procedure which was introduced in 1854 proved of great value for about one hundred years enabling the arbitrator to obtain the view of the court on difficult points of law before making his award and materially developing and enriching English commercial law in the process.

Increasingly, however, it became the subject of abuse to the point where England was becoming an unpopular forum for arbitration because a party who considered he was likely to lose could put off the evil day by calling for a Case to be Stated on some legal point or other, perhaps taking the result to the Court of Appeal. Even if he lost all the way, the benefits of delay could exceed the additional cost if enough were at stake while, in the meantime, he might be able to saddle his opponent with cash-flow problems serious enough to force him to settle unfavourably.

The power to set aside an award on the ground of an error of law or fact on its face was another possible source of abuse removed by the 1979 Act. It originated with the Writ of Certiorari in the Eighteenth century and originally suffered from the defect that the court had no power to vary as distinct from quashing an award, leaving the parties to start all over again.

In order to avoid this risk the practice grew up, especially in maritime arbitrations, of either giving reasons separately or not giving them at all.

The 1979 Act substituted two new but strictly limited ways of subordinating arbitrations to the court's views on legal issues, namely, judicial review, and the determination of preliminary points of law – see sections (e), and (b), *post*.

(b) Determination of a preliminary point of law

Under S2(1) of the 1979 Act any party, with the consent of the other side, or of an arbitrator (or umpire) who has entered on the reference, may apply to the court to determine a question of law arising during the course of the reference. Such an application must be made (and notice served) within 14 days of the giving of the relevant consent – 0.73. r.5(3). However, unless all parties support the application, the Court may not entertain it unless the question is one in respect of which leave to appeal under S1(3)(b) would be likely to be given *and* substantial savings of costs might result from its determination – S2(2).

In contrast to the position under S1, the court still has a discretion whether to give leave even if all the parties agree to it.

A preliminary question of law which, if rightly decided, would determine the whole dispute is an example of the wholly exceptional case where leave should be given.

Not only do the *Nema* guidelines discussed in section (g) on p. 96 apply but leave is likely to be granted less readily under S2 than under S1 – per Donaldson, LJ, in *Babanaft International Co. SA v Avanti Petroleum Inc.* [1982] 1 WLR 871 (CA) at pp. 882 and 883.

Lloyd, J, pointed out in *Vasso v Vasso* [1983] 3 All ER 211, that, provided the court and arbitrator are co-operative, the court's guidance on a point of law can be obtained during the course of an arbitration without the need to show a saving of costs; the arbitrator merely has to frame the question as an interim award.

No application may be made in regard to a question the determination of which is material to an award covered by an effective exclusion agreement between the parties – see S3(1)(c).

Little use has so far been made of S2; this may have been one of the purposes of the various restrictions.

(c) Appeals – reasons

A right to appeal may be nugatory in practice unless the reasons for the arbitrator's award can be extracted from him. S1(5) of the 1979 Act accordingly provides that, on the application of a party, either with the consent of the other side, or with leave of the court (which cannot be given if an exclusion agreement subsists), an arbitrator (or umpire) may be required to give, or amplify, his reasons for an award so that any question of law arising out of it can be considered.

However, where no reasons at all are given the court may not make an order even if the parties agree unless it is satisified that, before the award was made, one of them stated that a reasoned award would be required, or there is some special reason why this was not done – S1(6).

In considering whether to give leave under S1(5) the court should take into account that the arbitrator is not obliged to give reasons unless asked and should not be burdened with the task unless the reasons are likely to be needed for an appeal – *Trave Schiffahrtsgesellschaft mbH* v *Ninemia Maritime Corp.* [1986] 2 All ER 244. The court's jurisdiction to order further reasons under S(1)(5)(b) should be exercised very sparingly and only after giving the fullest consideration to whether leave to appeal is likely to be granted bearing in mind the *Nema* and *Antaios* guidelines – *Universal Petroleum Co. Ltd.* v *Handels-und-Transportgesellschaft mbH* [1987] 1 Ll. Rep. 517 (CA). Furthermore, the alleged uncertainty or ambiguity in the reasons given must arise out of the award not out of the arbitration generally, i.e. not out of matters not mentioned in the award itself.

(d) Appeal procedure

Directions as to the procedure which an appellant should adopt are set out in a Practice Note reported at [1985] 2 All ER 383.

In appropriate cases the appellant should ask the Court for an order staying execution of the award since a stay does not follow automatically from the fact of an appeal.

The Court may hear the substantive appeal at the time the application for leave is made, but does not normally do so.

Fairly severe time limits are imposed. Order 73 r.5 provides that applications for leave to appeal under S1(2) of the 1979 Act, or to remit or set aside an award under Ss22 or 23(2) of the 1950 Act, must be made and notice served within 21 days after the award has been made and published to the parties – see *Zermalt Holdings SA* v *Nu-Life Upholstery Repairs Ltd.* [1985] 2 EGLR 14. The same time limits apply to applications for reasons or further reasons except that where material reasons have been supplied after the publication the period runs from the date they were given.

(e) Appeals – the judicial review of awards

S1(3) enables any party not barred by an effective exclusion agreement to appeal on a question of law arising on an award[2] provided that he has either the consent of the other parties or leave of the Court.

Since the other side is likely to give consent in exceptional cases only, whether the court will give leave is usually crucial. In fact it may only do so if it considers that determining the question could substantially affect the rights of a party – S1(4) of the 1979 Act. Thus appeals on academic questions which were a feature of the old procedure are ruled out, while the court's power to impose conditions should be effective to obviate the earlier abuses. For instance, in *Mondial Trading Co. GmbH* v *Gill and Duffus Zuckerhandels GmbH* [1980] 2 Ll. Rep. 376, the applicant was required to pay into court the sum awarded plus two-thirds of the respondent's estimated appeal costs.

The next question, of course, is whether the court will in fact exercise its discretion to give leave to appeal. The statute does not lay down any specific criteria but the House of Lords filled the gap in *The Nema* by providng 'guidelines' which, together with a little subsequent 'fine tuning', will be found in sections (g) and (h) on pp. 96–9.

It will be appreciated that once the 'guidelines' are satisfied and leave to appeal is granted the issues will be determined in the normal way, that is to say, on the basis of whatever on balance is the correct solution; questions raised under the guidelines such as whether the arbitrator was obviously wrong are no longer relevant.

Upon determination of an appeal the court may either confirm, vary or set aside the award, or remit it to the arbitrator or umpire for reconsideration in the light of its opinion on the relevant point of law. The court's power to vary an award takes effect as if the award were that of the arbitrator or umpire except that it is no longer open to this appeal procedure – see S1(8).

Where an award is remitted, a fresh one must be made within three months unless the order otherwise directs – S1(2).

(f) Appeals from the High Court

Two kinds of appeal must be distinguished.

The first is from semi-administrative decisions by the High Court such as granting or refusing applications for leave to appeal or to determine a preliminary question of law where not all the parties consent, and applications for an order directing the arbitrator to provide reasons – see Ss1(3)(b), 2(1)(a) and 1(5)(b) respectively. These decisions are not appealable unless the High Court gives consent – see Ss1(6A) and 2(2A) of the 1979 Act added by S148(2)(3) of the Supreme Court Act 1981; without the consent the Court of Appeal will have no jurisdiction to hear the appeal – even apparently if it could be shown that the judge did not exercise his discretion judicially – *Aden Refinery Co. Ltd.* v *Ugland Management Co.* [1986] 3 All ER 737 (CA).

The second type of appeal is one on the merits, i.e. on the question of law itself. Here, either the High Court or the Court of Appeal must give leave, and the High Court must also certify the question of law to be either of general public importance or one which for some special reason the Court of Appeal should consider – see Ss1(7) and 2(3). Without that certificate, the Court of Appeal has no jurisdiction – *National Westminster Bank PLC* v *Arthur Young McClelland Moores & Co.* [1985] 2 All ER 817, (CA).

As to whether the judge *should* exercise his discretion in favour of leave to appeal, see below.

Where an appeal does go to the Court of Appeal it is customary for at least one member of the court to be a commercial judge.

There are no special provisions in arbitration cases governing appeals from the Court of Appeal to the House of Lords; the normal rules apply.

(g) The granting of leave to appeal – the *Nema* 'guidelines'

In *Pioneer Shipping Ltd.* v *BTP Tioxide Ltd. The Nema* [1982] AC 724, authoritative guidelines were laid down on the approach the courts should adopt in exercising their discretion whether to give leave to appeal. They are based on the view that Parliament's primary intention was to promote finality in awards and are as follows.

In 'one-off' cases leave should not be granted unless the Judge, after a mere perusal of the reasoned award, and without the benefit of adversarial argument, considers the arbitrator's decision obviously wrong; otherwise, that is to say, normally, the parties should be left, for better for worse, with the decision of the tribunal they have chosen.

If an applicant for leave to appeal considers his is *not* a 'one-off' case he should lodge an affidavit to that effect setting out the facts on which he relies; if the respondent disagrees, he should file an affidavit to the contrary.

At the other end of the scale lies the construction of standard contractual terms. Notwithstanding the public interest in the legal position being certain, leave should not be given unless there is a strong *prima facie* case that the arbitrator reached the wrong conclusion – indeed, if the issue only arose because of 'one-off' facts, the stricter approach mentioned above should be adopted.

Somewhere in the middle are cases where similar transactions are likely to recur. Lord Diplock thought that in such cases the judge would be justified in giving leave in the interest of uniformity if the decision seemed incorrect even though not so wrong that no reasonable person would have reached it. He gave as an instance a dispute as to the date when contracts were frustrated as a result of the Iran–Iraq war. In fact, after two or three cases with differing results, leave to appeal on this point was given in *The Wenjiang* [1982] 1 Ll. R. 128 (CA). It would, of course, still be necessary for the judge to be satisfied under S1(4) of the 1979 Act that the issue could substantially affect the rights of a party – see para (e) *ante*.

The House of Lords unanimously re-affirmed these guidelines in *Antaios Cia v Salen Rederierna* [1984] 3 All ER 229, Lord Diplock adding (at pp. 232-3) that they were not intended to be all-embracing or immutable. He also said that the hearing of an application for leave to appeal should be short and, perhaps regrettably, that reasons should not normally be given.

The Nema guidelines were applied in *Italmare Shipping Co. v Ocean Tanker Co. Inc. The Rio Sun* [1981] 2 Ll. R. 489, though Lord Denning, MR, expressed the view that the Court has a complete discretion outside the rules which the statute has laid down and that, being only 'guidelines', they could be stepped over without harm.

In *National Rumour Company SA v Lloyd-Libra Navegacao SA* [1982] 1 Ll. Rep. 472, the arbitrator was considered *prima facie* wrong but leave to appeal was refused because the parties had arranged for the arbitration to be a quick one.

In *Higgs and Hill Building Ltd. v University of London* (1983) 24 BLR 139, Parker, J, expressed the view that the guidelines were not to be applied rigidly but that the degree of flexibility did not differ between different classes of case.

The question of when the judge should give leave to appeal to the Court of Appeal under S1(6A) of the 1979 Act from his decision on an application to appeal was also dealt with in the *Antaios* case. It was decided that it should be given, and given only, where the then current guidelines as laid down by appellate courts seemed to require amplification, clarification or adaptation to changing practices.

In general, leave to appeal to the Court of Appeal is given less readily than appeal to the High Court – per Donaldson, LJ, in *Babanaft International SA* v *Avanti Petroleum Inc.* [1982] 1 WLR 871, at p. 882.

(h) Later modifications of the 'guidelines'

Notwithstanding that the 'guidelines' had recently been re-inforced in the *Antaios* and other cases, Browne-Wilkinson V-C decided not to apply them in *Lucas Industries PLC* v *Welsh Development Agency* [1986] 2 All ER 858. He was left in real doubt as to whether the arbitrator was right in law in an award under a rent review clause but gave leave because the decision would constitute an issue estoppel and also govern future rent reviews. A subsidiary reason was that a similar point would be likely to arise in other rent review cases.

The guidelines were also not applied where a new point arose which was potentially of great importance both to the parties and generally. In *Bulk Oil (Zug) AG* v *Sun International Ltd.* [1984] 1 All ER 386, the defendant had refused to load a consignment of UK crude to Israel on the ground that it would have been against UK policy. The plaintiff contended that this did not exonerate the breach because any such policy, if it existed, would have been void under EEC law. The arbitrator decided Bulk had been wrong to nominate Israel and the judge considered this decision right. Nevertheless, he granted leave to appeal against the interim award and also from his own judgment. There were no authorities precisely in point and the Court of Appeal thought that the views of both arbitrator and judge could well be wrong; accordingly, the judge's decision was that the matter should be referred to the ECJ – which incidentally, disagreed with both arbitrator and judge in due course.

In fact it seems that where it is important for authoritative guidance to be given as soon as possible the fact that a point is even capable of serious argument is sufficient to warrant the giving of leave.

A view rather less strict than the original guidelines was also expressed in *Aden Refinery Co.* v *Ugland Management Co.* [1986] 3 All ER 737 (CA), namely, that in appropriate cases leave should be given to appeal to the High Court where there have been conflicting decisions at first instance – not mere dicta – when leave to appeal to the Court of Appeal together with a certificate under S1(7)(b) should usually follow to enable the matter to be cleared up. Indeed, leave to appeal at least to the High Court should normally be given if there are conflicting decisions by arbitrators.

4. Exclusion agreements

An exclusion agreement is a written agreement between parties to a reference which excludes a right of appeal under S1 of the 1979 Act in relation to that award or, in the case of an application under S2(1)(a), which excludes an appeal in relation to an award to which determination of the preliminary question of law is material – see S3(1) of the 1979 Act.

It is of no effect in relation to a statutory arbitration – subsection 3(5) *ibid*.

In *Arab African Energy Corp* v *Olieprodukten Nederland BV* [1983] 2 Ll. Rep. 419, it was held that adopting the ICC Rules amounted to entering into an exclusion agreement in view of the waiver of the right to appeal which they contain.

An exclusion agreement need not form part of an arbitration agreement. Further, it may relate to a particular award, to awards under a particular reference, or to any other description of awards whether arising under the same reference or not – S3(2) of the 1979 Act. Subsection 3(4) renders ineffective any attempt by means other than an exclusion agreement to limit the jurisdiction of, or access to, the court, or to restrict the making of reasoned awards.

The existence of an exclusion agreement does not prevent the court from entertaining an application to appeal later, provided all the parties consent.

In the case of a domestic arbitration agreement an exclusion agreement is only effective if it is entered into after commencement of the arbitration – S3(6). The idea is that the respective strengths of the parties are often very different and it should not be possible to pressurise the weaker side into relinquishing its right to a review by the courts until it is in a position to weigh the possible consequences. A 'domestic' arbitration agreement for this purpose is an arbitration agreement which (a) does not expressly or by implication provide for arbitration outside the UK and (b) did not have as a party when it was entered into *either* (i) an individual who was a national of or habitually resident in a State other than the UK, *or* (ii) a body corporate either incorporated, or with its central management and control exercised, in such a State – S3(7).

Unless either the exclusion agreement is entered into after commencement of the relevant arbitration, or the award or question relates to a contract expressed to be governed otherwise than by English law, it has no effect where the reference relates in whole or in part to a question falling within the Admiralty jurisdiction of the High Court or where it relates to a dispute arising out of an insurance contract or to a 'commodity' contract (as defined) – S4(1). A 'commodity' contract for this purpose is defined by the Arbitration (Commodity Contracts) Order, 1979, as one for the sale of goods which are regularly dealt in on any of the commodity markets it mentions being a contract which is subject to the arbitration rules of any of the various bodies listed in the Order – see *post.* at pp. 170-1.

It seems that the purpose of limiting the use of exclusion agreements in this way is to increase the number of issues of general applicability which will continue to reach the courts and enrich English commercial law. The Secretary of State has wide powers to vary the three classes by order – S4(3).

Where the parties to a non-domestic arbitration agreement have entered into an exclusion agreement and the alleged fraud of a party is an issue, the Court has no power to intervene under S24(2) of the 1950 Act unless the exclusion agreement otherwise provides – S3(3) of the 1979 Act.

5. Interventions under the 1950 Act

So far this Chapter has been concerned almost exclusively with appeals and references to the Courts under the 1979 Act. Before mentioning court intervention under the 1950 Act, it is convenient to contrast remitting an award with setting it aside. The main difference is that in the former case the arbitrator must consider, or re-consider, the matters to which the Court has drawn his attention, while if an award is set aside that arbitration is at an end and the arbitrator is *functus officio*. The arbitration agreement would, however, normally remain effective; indeed as pointed out at p. 8 the court setting the award aside may order the time taken up by the abortive arbitration to be disregarded for limitation purposes Limitation Act 1980, s34(5).

Remission may be appropriate where the arbitrator has committed merely a mistake, where his 'misconduct' is of a technical nature and does not indicate any degree of dishonesty. Setting aside is usually reserved for more serious cases. Procedural lapses which do not result in unfairness will not cause an award to be set aside – *Overseas Fortune Shipping Pte. Ltd.* v *Great Eastern Shipping Co. Ltd.* [1987] 1 Ll. Rep. 270.

The time limit for applications to remit or set aside has been mentioned on p. 94. The modern practice is to serve notice of the proceedings on the arbitrator where misconduct is alleged who may then either appear, file an affidavit or ignore it. In the absence of fraud or quite exceptional circumstances costs would not be awarded against the arbitrator.

If the good parts of an award can be separated from the bad, the remitter for re-consideration may be confined to the letter.

(a) Wherever there is a reference to arbitration the High Court has power to remit the matters referred for reconsideration – S22 of the 1950 Act. The arbitrator (or umpire) must make his reconsidered award within three months unless the order otherwise directs. The Section must now be read subject to S1(1) of the 1979 Act which, *inter alia,* removed the Court's jurisdiction to remit an award on the ground of errors of fact or law on its face, and it is not quite clear what S22 is now intended to cover. However, the discovery of new evidence is one ground for remitter which presumably still survives. To persuade the Court to exercise its power in such a case it will usually be necessary to show, first, that at the time of the arbitration the evidence was unavailable and could not have been obtained with reasonable diligence, second, that it would be likely to have had a substantial effect on the findings and third, in appropriate cases, that there was no opportunity to ask the arbitrator to delay making his award while the possibility of obtaining evidence of the type in question was considered – *Whitehall Shipping Co.* v *Kompas Schiffahrtskontor GmbH, The Stainless Patriot* [1979] 1 Ll. Rep. 589. The court may also use S22 to remit an award to the arbitrator for correction where its operative part is inconsistent or ambiguous – see *Moran* v *Lloyd's* [1983] 2 All ER 200 (CA). Inconsistency in reasoning, however, must now be accepted by the parties if it proceeds from an error of fact, though if it indicates an error of law it can be appealed subject to the provisions of S1 of the 1979 Act and the normal guidelines.

(b) Where an arbitrator or umpire has misconducted[3] either himself or the proceedings, he may be removed and his award set aside; the latter may also occur if an arbitration or award has been improperly procured. The Court may order any money made payable by an award to be brought into Court or otherwise secured pending determination of an application to set the award aside – S23 of the 1950 Act.

(c) Where an arbitration agreement relates to future disputes and the fraud of a party is an issue, the court has power to order that the arbitration agreement is to cease to have effect, to revoke the authority of any arbitrator or umpire appointed under it and, by way of corollary, to refuse a stay of any action brought in breach of the agreement – S24(3).

(d) An arbitrator has no inherent power to order a party to give security for costs and the arbitration agreement is unlikely to confer it. However, the High Court has power under S12(6)(a) of the 1950 Act to make such an order and will exercise it in the same way as it would in an action, the most common instance being where the claimant – or a defendant who makes a substantial counterclaim – is a foreign corporation without adequate assets in this country – see *Samuel J. Cohl & Co.* v *Eastern Mediterranean Maritime Ltd.* [1980] 1 Ll. Rep 371.

6. The courts' 'inherent jurisdiction' to supervise arbitrations

The powers which the court has under its inherent jurisdiction are those needed for the protection of legal or equitable rights, such as the right to have a dispute decided by an unbiased arbitrator. However, despite considerable controversy it seems that at common law there is no general power to supervise arbitral proceedings as would be the case with inferior courts and with tribunals which are not consensual, e.g., statutory arbitrations – see per Lord Diplock in *Bremer Vulkan Schiffbau und Maschinenfabrik* v *S India Shipping Corporation* [1981] AC 909 at pp. 983-4. Similarly, no injunction can be obtained other than for the enforcement or protection of a legal or equitable right – *ibid* at p. 980.

Such powers as the court has are apparently confined to intervening where real, as distinct from technical, misconduct arises. Thus Lord Scarman said, *ibid.,* at p. 997 that subject to ouster of its jurisdiction by an exclusion agreement, 'the courts will act to prevent injustice arising in arbitration proceedings where it is necessary so to do' but this does not extend to granting an injunction against proceeding with an arbitration even though the delay in prosecuting it has been such that a fair trial would no longer be possible. Each party owes the other an obligation to join in asking the arbitrator to terminate a serious delay. The *Bremer Vulkan* case certainly seems to illustrate the modern tendency to leave it to the parties and their arbitral tribunal to get on with it, or, to put it another way, to place more weight upon finality and certainty than upon legal precision.

Where the Court would have jurisdiction apart from an arbitration clause it retains a residual jurisdiction even though a *Scott* v *Avery* provision may defer its exercise; full jurisdiction will be resumed in certain circumstances, for example, where a sole arbitrator is removed, or where the arbitration agreement is frustrated or ceases to have effect by virtue of an order under S25(2) of the 1950 Act.

The existence of an arbitration clause does not constitute an absolute bar to bringing an action but either entitles or obliges the Court to grant a stay if the other side relies on it in due time.

Notes

1. When applications are made in Chambers it is the responsibility of both parties to ensure that all relevant documents are lodged at least two days before the hearing so that the judge may read them – see Practice Note reported at [1985] 2 All ER 384.
2. The award must be on an 'arbitration agreement' as defined – see p. 25.
3. An early instance of what the English courts would certainly have regarded as misconduct occurred when Rome was asked to arbitrate in a border dispute between Gallic tribes. Neither wishing to appear unduly grasping, some gaps occurred between 'the boundaries' they claimed and Rome promptly annexed the land involved.

9 International arbitrations

1. General

A few notes on this vast subject are included mainly for the benefit of those contracting with persons and bodies resident abroad. In such cases, it is usually not for the advantages mentioned in Chapter 2, but in spite of the disadvantages, that arbitration is chosen as the means of resolving disputes. There is indeed no real alternative where neither party will contemplate using the Courts of the other. Where both parties are foreign it is quite common for the English courts to be chosen, perhaps as much because of the practical, non-academic approach and the comparative predictability of the law as for other reasons. On the other hand, United Kingdom parties seldom agree willingly to foreign courts, though whether this is for reasons of language, an unfamiliar legal system, delay, expense or undue insularity no doubt varies from one to another.

A party about to enter into a contract with a person or body resident abroad, and his advisers, should bear in mind that the civil law countries – which comprise most Continental countries and their former colonies – have many procedural and some substantive rules in common, while the common law countries – very broadly, England, the USA and most of the British Commonwealth – are in a comparable position in that they also share very similar legal traditions. So far as both arbitration law and procedure are concerned there has recently been an appreciable narrowing of the gap between the two systems, but they are still far enough apart for it to be desirable for each party to take what care it can not to be saddled with proceedings which are wholly alien.

2. The arbitration agreement

Where there is an international element it is usually advisable for the arbitration agreement to cover certain additional topics and to deal with others rather differently.

Number of arbitrators

It is common for there to be three arbitrators, or two arbitrators and an umpire, instead of the single arbitrator of most domestic arbitrations – and, incidentally, of many ICC arbitrations when less than $1m. is at stake. Usually each party nominates one arbitrator and these arbitrators then make the third appointment. Often the result in countries where there is no awareness of, or scant heed is paid to, the principle of the independence of arbitrators, such an appointee will consider himself the advocate of the party who appointed him. This is to be regretted but the harm is less where the position is understood from the beginning, and it may at least assist in reconciling a party to an arbitration which turns out unfavourably to know that his point of view was fully represented when the decision-making took place.

Where appointments are made by an international body such as the ICC care is usually taken that each of three arbitrators comes from a country different from that of any of the others or of either party.

Procedure

In an international arbitration neither party is likely to favour recourse to the courts – anywhere – except when it is unavoidable. There must, however, be some degree of order and it is therefore generally desirable that the arbitration should be subject to a degree of surveillance by some arbitral body and that it should be conducted in accordance with a set of rules which can be studied beforehand and approved or rejected as the case may be. Some possibilities are mentioned on pp. 106–7.

Other special terms

Apart from vital decisions as to the choice of arbitral body and the extent of its involvement, a number of matters special to international arbitrations require attention.

Often one party to a contract is initially in a position to make his views prevail but where this is not so the parties may have difficulty in concurring on the venue, or even as to the applicable law (or laws). Sometimes, all that the parties may be able to agree is that the arbitration shall not take place in, or be subject to the laws of, the country of the other side.

The identity of the law of the contract may of course have vital repercussions, but the venue is also very important since it will usually determine the lex fori, which in turn will govern much of the procedure – see Chapter 1 p. 9.

Care should be taken to specify the language in which the arbitration is to be conducted where this could be in doubt.

Failure to deal with any of the points set out under the last two headings at the time of the arbitration agreement may well be the cause of substantial delays and additional expense later on.

3. Applicable rules and administered arbitrations

(i) Where the parties are both from common law countries – indeed where they are not – consideration might well be given to the model arbitration clause set out in the 1985 Edition of the Rules of the London Court of International Arbitration. It is as follows:

'Any dispute arising out of or in connection with this contract, including any question regarding its existence, validity or termination, shall be referred to and finally resolved by arbitration under the Rules of the London Court of International Arbitration, which Rules are deemed to be incorporated by reference into this clause'.

Reference to this body or the use of its rules does not mean that the arbitration has to be carried out in England.

(ii) Similar considerations apply to the Chartered Institute of Arbitrators which has its own rules and is prominent in the promotion of international arbitration both in London and in its various overseas branches.

(iii) No doubt the body most frequently used for administrating international arbitrations – in spite of criticisms levelled against it of high costs and delay – is the International Chamber of Commerce in Paris. The expense element largely stems from a substantial administrative levy and the fact that ICC fees are payable in advance and based on a percentage of the amount at stake – though the latter principle has some advantages. As from July, 1986, there has been some modification but ICC arbitrators will still occasionally be paid too little, and more often too much for their services. The initial fee for registering a request for arbitration has in fact been raised from $500 per party to $2,000 payable by the claimant alone, but a ceiling of an additional $48,500 has been placed on the administrative charge – which many litigants would consider quite enough. The fee for the tribunal has been subject to slight mitigation in two ways. Only 50% of the total fee is demanded initially, the remainder being due before the Terms of Reference become effective and the tribunal actually begins the arbitration; second, where a party has paid cash for half what is due – including the share of a defaulting party – the other half may now be covered by a bank guarantee.

By way of contrast it may be noted that it costs only £200 to register an arbitration with the London Court of International Arbitration and thereafter its administrative costs are charged out at £50 per hour plus actual outgoings. Again, the fees charged by arbitrators in the UK usually depend on time spent, though complexity and importance are also taken into account. In an average case an arbitrator might charge £300–1,250 per day for hearings and meetings, and £60–250 per hour for time spent otherwise on the reference. Such fees are normally paid in arrears.

The main distinguishing features of ICC arbitrations are, first, arbitrators are appointed by 'national committees' who may lack practical up-to-date experience of arbitration, second, before the arbitration begins the parties and arbitrators negotiate and sign Terms of Reference which define the issues which are to be resolved and, third, awards are vetted by the ICC Court before being delivered to the parties.

(iv) Amongst other bodies administering international arbitrations are the American Arbitration Association and various regional bodies such as the Euro-Arab Chambers of Commerce. The International Centre for the Settlement of Investment Disputes should also be mentioned although, since it was formed under the aegis of the World Bank to deal with disputes between investors and Governments, arbitrations are not an everyday occurrence.

(v) Certain trade associations whose membership is international have their own arbitration systems which have regard to customs of the trade rather than the nationality or residence of their members. An example is the Federation of Oil, Seeds and Fats Associations, FOSFA.

In the case of some associations, English law is deemed to apply to their standard contracts.

(vi) There are various other sets of international arbitration rules, and indeed conciliation rules. Special mention may be made of the 1976 rules of UNCITRAL – the United Nations Commission on International Trade Law. In June 1985 UNCITRAL agreed a Model Law of arbitration with a view to its adoption as an arbitration code by individual countries with or without modifications. Canada has in fact taken advantage of these provisions.

4. Evidence

One of the most important areas in which civil law systems, which derive ultimately from Roman Law, differ from those of the common law is in regard to evidence. An attempt has been made by the International Bar Association in its IBA Rules of Evidence to produce a code of Rules which would form a bridge between the two, and it would be sensible wherever the potential parties come from opposing systems to consider whether they might not be acceptable to both. The importance of the differences which exist between the two systems may be gauged from the following isolated comments.

In some civil law countries evidence is not normally taken on oath; in some, litigants place particular emphasis on their *right* to produce evidence and may require leave of the court to do so; it may be considered improper to interview witnesses before the trial; a litigant may not be entitled to call witnesses in support of his own case; a person may only be entitled to give evidence *against* himself; the procedure may be to set out witnesses' statements in 'Articles of Proof'; and so on.

In the case of expert witnesses, civil law courts, perhaps reluctant to choose between two experts, will usually appoint a court expert and rely on his evidence, even though the parties produce experts of their own. The court appointee is not normally examined in chief nor subject to cross-examination.

There is no full discovery and applications in regard to specific documents are not very common; in Italy, for instance, one must prove not only that the document exists and that the other person has it, but provide some knowledge of its contents.

5. Enforcement of foreign awards

By reason of the over-lapping nature of the various international agreements, the precise position as regards the enforcement of foreign awards in England, and of English awards abroad, is rather confused.

Convention awards

In 1958 the Convention on the Recognition and Enforcement of Foreign Arbitral Awards, generally known as 'the New York Convention', was adopted by an international conference in that city. The United Kingdom acceded to it on 23 December 1975, and the Arbitration Act, 1975, as its preamble declares, was enacted to give it effect. It is convenient to begin with this Act since where it applies, that is to say, the award is a 'Convention award', Part II of the 1950 Act has no application. Nevertheless there is very little practical difference between the enforcement of Convention awards under the 1975 Act and of 'foreign awards' under Part II of the 1950 Act; the provisions of Ss3–5 of the 1975 Act resemble those of S36–9 of the 1950 Act.

Part II of the 1950 Act, and the Geneva agreements of the 1920s will be dealt with at pp. 110-1 *post*.

S7 of the 1975 Act defines a 'Convention award' as an award pursuant to a written arbitration agreement which is made in any party to the Convention (other than the United Kingdom). It is to be treated as binding for all purposes on the persons between whom it was made and is enforceable as such anywhere in the United Kingdom, including, in England, enforcement by action under s26 of the 1950 Act and reliance on it by way of defence, set off or otherwise.

The critical time for deciding whether an award is a 'Convention award' is the time when proceedings are taken to enforce it – see *Kuwait Minister of Public Works* v *Sir Frederick Snow* [1984] 1 All ER 733 (HL) in relation to a 1973 award; Kuwait became a party to the Convention in 1978.

Mention in an Order in Council made for the purpose provides conclusive evidence that a particular State is a party to the Convention – S7(2) of the 1975 Act. The current list according to the Arbitration (Foreign Awards) Order, 1984, as amended by SI 1985/455, SI 1986/949 and SI 1987/1029 is set out at pp. 179-80.

Enforcement of a Convention award may be refused under S5(2)(3) of the 1975 Act if (i) it would be contrary to public policy to enforce it, (ii) the award relates to a matter not capable of being settled by arbitration, or (iii) the party against whom it is invoked shows that it was subject to any of the following procedural defects, namely:

(a) a party to the arbitration agreement was under some legal incapacity;
(b) the arbitration agreement was not valid under the law to which the parties made it subject, or, if they gave no indication on the matter, under the law of the country where the award was made;
(c) the applicant was not given proper notice of the appointment of the arbitrator, or of the hearing, or was otherwise unable to present his case;
(d) the award contains decisions on matters outside the reference which cannot be severed from it;
(e) the composition of the tribunal, or the procedure, was not in accordance with the parties' agreement or, failing such agreement, with the law of the country where the arbitration took place; or
(f) the award has not become binding, or has been set aside or suspended.

If proceedings to set the award aside are pending in the country where it was made, the court may adjourn an application to enforce it but may require the other party to give security – S5(5) of the 1975 Act.

Awards under the Geneva Convention

Where an award made overseas is not a Convention award as defined above, for example, because it was made in a country which is not a party to the Convention, it may well be enforceable as a 'foreign award' under Part II of the 1950 Act, which now covers the obligations arising under the Geneva Convention of 1927.

To constitute a 'foreign award' for this purpose it must satisfy the following criteria:

(a) it must not qualify as a Convention award;
(b) it must have been made pursuant to an arbitration agreement to which the Geneva Protocol of 1923 applies; and
(c) each party must have been subject to the jurisdiction of a Power which an Order in Council states is a party to the Geneva Convention, 'reciprocal provisions' having been made – S35 of the 1950 Act.

NB There is no need for reciprocal provisions in the case of 1958 Convention countries. The relevant list appears in the Arbitration (Foreign Awards) Order, 1984, and as amended is set out at p. 179.

Signatories to the 1923 Geneva Protocol *inter alia* recognised the validity of arbitration agreements made between parties of different States whatever the situs of the arbitration and that the procedure should be governed by the will of the parties and the *lex fori*. Each signatory State undertook to enforce such agreements (i.e. ensure that a dispute is referred to arbitration if a party so applies) and also to execute awards made in its own territory and in accordance with its laws.

The 1927 Geneva Convention went rather further. Thus it bound each signatory to enforce in accordance with its own procedure awards made in the territory of any signatory between parties who were subject to the jurisdiction of any of them, provided that recognition or enforcement of the award was not contrary either to public policy or the principles of law of the signatory concerned.

For a 'foreign award' to be enforced, it must:

(a) have been made pursuant to an arbitration agreement which was valid under the law by which it was governed;
(b) have been made by the tribunal provided for in the agreement or as agreed by the parties;
(c) have conformed with the law governing the arbitration proceedings;

(d) have become final where made, which by virtue of S39 will not be treated as the case if proceedings for contesting the validity of the award are pending; and

(e) have related to a matter which may lawfully be referred to arbitration in England – S37(1) of the 1950 Act.

Moreover, the enforcement must not be contrary to public policy or the law of England – *ibid*.

S37(2) goes on to declare a 'foreign award' unenforceable if it is annulled in the country in which it is made, or it does not deal with all the matters referred or goes beyond them, or if the other party was not given sufficient time to present his case or was under some legal incapacity and not properly represented. However, if the award does not deal with all the matters referred the court may postpone its enforcement or require security to be given by the party seeking to enforce it – S37(2). The court also has power to postpone or refuse enforcement in certain cases where a party has grounds for attacking the validity of an award – see S37(3).

The evidence needed to enforce a foreign award is set out in S38 of the 1950 Act.

If all these provisions are satisfied, the foreign award may be enforced either by action or in the same manner as a domestic award under S26 of the 1950 Act and may be relied on by way of set off or defence in any proceedings in England – S36.

It is to be noted that, as a matter of international comity, the High Court has recognised the validity of a tribunal set up by the USA and Iran to determine disputes between their nationals exclusively; an action brought by a party whose claim before that tribunal had failed was consequently struck out – *Wallal* v *Bank Millat* [1986] 1 All ER 239.

An award under which a foreign arbitrator failed to apply US anti-trust law could not be enforced in the US as to do so would be against public policy – see Article 5(a) of the New York Convention, and compare S5(3) of the 1975 Act. However, it seems that the US Supreme Court is resolving in favour of arbitration previous doubts as to whether it is an appropriate tribunal for determining cases in which triple-damages can be awarded – see *Mitsubishi* v *Soler Chrysler – Plymouth* (1985) 87 L. Ed. 2d. 444, an anti-trust case, and, on 7 June 1987, *Shearson American Express* v *McMahon* (a 'RICO' case).

Appendix 1(a)

IN THE MATTER OF THE ARBITRATION ACTS 1950-79
IN THE MATTER OF AN ARBITRATION
BETWEEN:

                JOHN HARRISON   Claimant

                        and

              RAYMOND STANLEY   Respondent

DIRECTIONS

Upon hearing Mr Arthur James of John James and Co. of the Strand solicitors to the Claimant and Dr Philip Halstead of Pollards Hill, solicitor to the Respondent, I direct as follows:

1   That the Claimant's solicitors do deliver Points of Claim to the Respondent's solicitors with a copy to me on or before the   1987.

2   That the Respondent's solicitors do deliver Points of Defence together with Points of Counterclaim with a copy to me within 28 days of receipt of the Points of Claim.

3   That the Claimant's solicitors do deliver to the Respondent's solicitors with a copy to me their Reply and their Points of Defence to Counteraction within 28 days of receipt of the Points of Defence.

4   That the Respondent's solicitors do deliver any Reply to the Defence to Counterclaim to the Claimant's solicitors with a copy to me within 14 days of receipt of such Points of Defence thereto.

5   That within 14 days of closure of the pleadings each side do deliver to the other a list of all documents in their power or under their control relevant to the issues in this arbitration but excluding any correspondence (a) before . . . and (b) otherwise than with . . ., inspection to be seven days thereafter.

6   That two agreed bundles be prepared bundle (A) to include only such documents as either party thinks I should read before the hearing and a copy to be delivered to me seven days before the hearing, all other documents to be in bundle (B), both bundles to be in chronological order and numbered consecutively.

7   That both parties provide me with short written statements of their proposed submissions seven days before the hearing.

8   That the parties consult with each other one month before the provisional date of the hearing as to any changes in the estimate of its length viz. 2 days.

9   That the Claimant be responsible for making all administration arrangements for the hearing provisionally fixed for the

Costs in the arbitration
Liability to apply

Dated this   day of   Arbitrator
1987

Appendix 1(b)

SCOTT SCHEDULE

1. No. of Item	2. Client on whose account Claimant performed services	3. Claimant's Comments	4. Value of Services claimed	5. Amount offered by Respondent	6. Respondent's Comments	7. Arbitrator's Remarks
1.	A. Jones	Claimant introduced this client and was present at first interview	£200	£100	Apart from attending a short introductory meeting the Claimant did nothing for this client	
2.	C. Evans	Claimant looked after this client throughout	£500	£500	Agreed	
3.	R. Smith	Claimant in charge of all work and was only assisted on two occasions (10th January/15th March) by other members of staff	£900 (90% of total fee charged)	Nil	Client complained that claimant was inefficient and rude and has so far refused to pay.	
	etc.	etc.		etc.		

Appendix 1(c)

Peremptory Notice

I appoint the . . . day of   next at 2.30 pm at   for proceeding with this arbitration and I give notice that in case either party shall fail to attend without having previously shown me good and sufficient cause I shall if the other party being present so requests proceed with the reference *ex parte*.

Dated this   day of

                                                                 Arbitrator

Appendix 1(d)

IN THE MATTER OF THE ARBITRATION ACTS 1950–1979

IN THE MATTER OF AN ARBITRATION

BETWEEN:

BOA CONSTRICTION CO. PLC   <u>Claimant</u>

and

THE OMNISCIENT MERCHANT BANK PLC   <u>Respondent</u>

AWARD

1. The Claimant company which was desirous of acquiring a Delaware corporation called Rabbits Inc. entered into a contract with the Defendant company dated 1st January, 1986, to advise them in regard to a bid. The said contract contained an arbitration clause in the following terms: 'Any dispute arising out of this contract shall be referred to an arbitrator agreed between the parties' due provision being made in the event of their failing to agree.

2. Disputes arose and the parties by letter dated 15 January 1987, jointly appointed me as arbitrator informing me of their agreement that the arbitration should be conducted in accordance with the Rules of the Chartered Institute of Arbitrators. I accepted the appointment by letter dated the 18 January 1987.

3. The Claimant's main contentions as set out in the correspondence and later in the pleadings were (i) that senior officers of the Respondent bank had taken advantage of their knowledge of the forthcoming bid to purchase a substantial number of shares in Rabbits Inc. on the NY Stock Exchange thereby inflating their price to a point where the Claimant had to increase

its bid; and (ii) that the Respondent had negligently advised the Claimant to hedge against a rise in the dollar in terms of sterling to cover the dollar loan which the Claimant raised to make the acquisition ('the hedging transaction').

4. The Respondent (a) admitted the improper purchases of shares but denied that the damage was as alleged; (b) denied having given negligent advice as alleged; and (c) counterclaimed for damages by reason of an allegedly libellous letter from the Claimant to the Press which contained disparaging remarks at the Respondent's expense.

5. I held a preliminary meeting on 10 February 1987 at the Arbitration Room, Gray's Inn. Both the parties having previously expressed a desire to be represented by leading Counsel in view of the importance of the matters to be discussed, Mr White, QC, instructed by Messrs Plonque and Rosay, appeared for the Claimant and Mr Black, QC, instructed by Messrs Dodson and Fogg appeared for the Respondent.

6. At the said meeting two preliminary issues were raised:
(1) The Claimant contended that the counterclaim for damages for libel was outside the terms of the arbitration agreement. After hearing arguments and the citation of various authorities I concluded that I had no jurisdiction to deal with this aspect of the matter.
(2) The Respondent contended that the arbitration should be consolidated with an arbitration in which I had been appointed arbitrator and which was shortly to commence between the Respondent and Advisory Bank Inc., a New York bank which had in fact supplied advice in regard to the hedging transaction. The Respondent produced an affidavit from Advisory Bank Inc. consenting to the consolidation and asserted that in a conversation between the Respondent's Chairman and the Claimant's Managing Director the latter had agreed to such consolidation on the grounds that it would save time and trouble. It was stated on behalf of the Claimant that its Board had since considered the suggestion formally and had rejected it on the ground that it would involve delay and additional expense.

7. I considered the consolidation proposal on the assumption that the stated facts would be duly established in the light of the jurisdiction given me by the said Rules, Rule C(i) of the Schedule to which enables me to allow other parties to be joined with their express consent but only after hearing representations from the parties. I was informed that Counsel instructed to appear for Advisory Bank Inc. wished to address me but I rejected the invitation on the ground that unless and until I agreed to consolidation he had no locus standi.

8. The Claimant conceded that consolidation would in some ways be a convenient course but argued that it would be prejudicial by reason of the additional cost and delay arising partly but not exclusively from the fact that the proposed additional party does not carry on business in this country. I accepted this argument and since I do not consider the Claimant in any way bound by its Managing Director's expression of intention I reject this proposal. At the conclusion of the meeting I made directions in the present arbitration.

9. The hearing which also took place at Gray's Inn began on 5 August 1987 and continued on the two following days.

10. The background facts of the main issue were not in dispute and may be summarised as follows. The Respondents originally recommended the Claimant to offer $1 each for the 20 million shares of no-par value in Rabbits Inc. which comprised the whole of its issued share capital and were then standing at 65 cents. Before a formal offer could be made, however, the price rose rapidly to 95 cents and the Respondent consequently recommended that the $1 offer should be raised to $1.30. This offer was in due course accepted in respect of 18 million shares. It was later admitted that key members of the Respondent's staff had been purchasing substantial blocks of the shares.

11. The claimant contended that the damage to it on these facts amounted to 30 cents per share, a total of $5.4m and that this should be expressed in sterling at the rate obtaining on 15 April 1986 when payment for the Rabbits shares was made (agreed to be $1.20 to the £).

12. The Respondent contended (i) that the rise from 65 to 95 cents could not be wholly attributed to the insider-dealing since the Stock Exchange was buoyant at the time; (ii) that it was reasonable to assume that had only $1 been offered acceptances would have been less and the proportion acquired might well have fallen below 80% (instead of the actual 90%) in which event the holding would have been materially less valuable in view of certain provisions of US tax law; and (iii) that in any case any damage found should be expressed in dollars; since the contract between the parties related to a purchase in US dollars that was the currency with which it had the most real connection or, to put it another way, a loss in dollars should be made good in dollars.

13. The background facts relating to the hedging transaction are similarly not in dispute. In order to finance the transaction the Claimant borrowed $12m. for one year, which the Respondent advised should be made the subject of a swop transaction. However, the sterling/dollar rate moved from 1.20 to 1.50 during the relevant year and the 'swop' had the effect of wiping

out the £2m. profit which the Claimant would otherwise have made on repayment of the loan. The Claimant accordingly claimed £2m. plus costs agreed to be in the sum of £100,000.

14. The Respondent did not dispute that the bulk of informed opinion at the time was that the dollar was more likely to fall than rise but maintained that that was not the universal view and that advising a course which would prevent the Claimant, which was an ordinary trading company, from incurring a loss was the correct advice in the absence of a clear indication that the Claimant wished to embark on a currency speculation.

15. The Claimant called Mr Pooh Baer and the Respondent Mr Hocus Z. Blank, both experienced US investment bankers, to give expert evidence in regard to (a) market conditions at the relevant time, (b) as to what the likely effect of a lower bid would have been, and (c) as to market sentiment in regard to exchange rates at the time it was decided to enter into the hedging transaction.

16. Having considered this evidence and the arguments of Counsel I reached the following conclusions, namely –

(a) that there was some substance in the first of the Respondent's arguments set out in paragraph 12 above and that one-third of the rise, namely, 10 cents should be attributed to reasons other than the insider-dealing;

(b) that there was no reason to think that the Respondent's original advice to bid $1.00 would have been either more or less attractive in the conditions then obtaining than its advice to bid $1.30 later on, or that it would have resulted in a different proportion of acceptances;

(c) that since the parties both carried on business in the United Kingdom, the contract was drawn in London and the Respondent's fee under it was payable in sterling, sterling was the currency with which the contract had its most real connection. Further, that in view of the recent sharp appreciation in the dollar an award in dollars would not lead to a just result;

(d) that the Claimant knew the purpose of the swop transaction perfectly well, and accepted it, and that if it had any desire to speculate against the dollar that desire was not communicated to the Respondent.

17. In the result, I HOLD
   (i) that the damage to the Claimant under this head was $3.6m. which at the rate of $1.20 to the £ amounts to £3m.
   (ii) that the claim that the Respondent was negligent in its advice in regard to the hedging transaction fails.

AND I AWARD and DIRECT that in full and find settlement of all claims raised herein, the Respondent do pay to the Claimant forthwith the sum of

£3m. with interest at 12 per cent per annum from 15 April 1986 to the date hereof.

18. A separate hearing in regard to the question of costs took place at the same venue on 15 August, 1987.

19. The Respondent having been unsuccessful on both the preliminary points I AWARD and DIRECT that he pay the costs attributable thereto which I tax and settle in the sum of £12,000, including VAT.

20. The Claimant succeeded on one of the two main points though he failed on one aspect of it which reduced his claim by one third; he failed completely on the second main issue which took up merely half the time.

21. Accordingly, I further AWARD and DIRECT that the Respondent further pay to the Claimant the costs of this my AWARD which I tax and settle in the sum of £8,000 plus VAT of £1,200. I make no further directions as to costs.

22. I further DIRECT that should the Claimant take up this AWARD the Respondent re-imburse him the said sum of £8,000 plus VAT of £1,200.

Witnessed by
.... of
....
....
....

(Signed)

ARBITRATOR
London 18 August 1987

Appendix 2

ARBITRATION ACT 1950

(1950, c. 53)

An Act to consolidate the Arbitration Acts 1889 & 1934 [28th July 1950]

PART I

GENERAL PROVISIONS AS TO ARBITRATION

Effect of Arbitration Agreements, etc.

Authority of arbitrators and umpires to be irrevocable
1. - The authority of an arbitrator or umpire appointed by or by virtue of an arbitration agreement shall, unless a contrary intention is expressed in the agreement, be irrevocable except by leave of the High Court or a judge thereof.

Death of party
2. - (1) An arbitration agreement shall not be discharged by the death of any party thereto, either as respect the deceased or any other party, but shall in such an event be enforceable by or against the personal representative of the deceased.

(2) The authority of an arbitrator shall not be revoked by the death of any party by whom he was appointed.

(3) Nothing in this section shall be taken to affect the operation of any enactment or rule of law by virtue of which any right of action is extinguished by the death of a person.

Bankruptcy
3. - (1) Where it is provided by a term in a contract to which a bankrupt is a party that any differences arising thereout or in connection therewith shall be referred to arbitration, the said term shall, if the trustee in bankruptcy adopts the contract, be enforceable by or against him so far as relates to any such differences.

(2) Where a person who has been adjudged bankrupt had, before the commencement of the bankruptcy, become a party to an arbitration agreement, and any matter to which the agreement applies requires to be determined in connection with or for the purposes of the bankruptcy proceedings, then, if the case is one to which subsection (1) of this section does not apply, any other party to the agreement or, with the consent of the committee of inspection, the trustee in bankruptcy, may apply to the court having jurisdiction in the bankruptcy proceedings for an order directing that the matter in question shall be referred to arbitration in accordance with the agreement, and that court may, if it is of opinion that, having regard to all the circumstances of the case, the matter ought to be determined by arbitration, make an order accordingly.

Staying court proceedings where there is submission to arbitration

4. - (1) If any party to an arbitration agreement, or any person claiming through or under him, commences any legal proceedings in any court against any other party to the agreement, or any person claiming through or under him, in respect of any matter agreed to be referred, any party to those legal proceedings may at any time after appearance, and before delivering any pleadings or taking any other steps in the proceedings, apply to that court to stay the proceedings, and that court or a judge thereof, if satisfied that there is no sufficient reason why the matter should not be referred in accordance with the agreement, and that the applicant was, at the time when the proceedings were commenced, and still remains, ready and willing to do all things necessary to the proper conduct of the arbitration, may make an order staying the proceedings.

(2) [*Repealed by Arbitration Act 1975, s.8(2)(a)*]

Reference of interpleader issues to arbitration

5. Where relief by way of interpleader is granted and it appears to the High Court that the claims in question are matters to which an arbitration agreement, to which the claimants are parties, applies, the High Court may direct the issue between the claimants to be determined in accordance with the agreement.

Arbitrators and Umpires

When reference is to a single arbitrator

6. Unless a contrary intention is expressed therein, every arbitration agreement shall, if no other mode of reference is provided, be deemed to include a provision that the reference shall be to a single arbitrator.

Power of parties in certain cases to supply vacancy
7. Where an arbitration agreement provides that the reference shall be two arbitrators, one to be appointed by each party, then, unless a contrary intention is expressed therein –

(*a*) if either of the appointed arbitrators refuses to act, or is incapable of acting, or dies, the party who appointed him may appoint a new arbitrator in his place;

(*b*) if, on such a reference, one party fails to appoint an arbitrator, either originally, or by way of substitution as aforesaid, for seven clear days after the other party, having appointed his arbitrator, has served the party making default with notice to make the appointment, the party who has appointed an arbitrator may appoint that arbitrator to act as sole arbitrator in the reference and his award shall be binding on both parties as if he had been appointed by consent:

Provided that the High Court or a judge thereof may set aside any appointment made in pursuance of this section.

Umpires
8. – (1) Unless a contrary intention is expressed therein, every arbitration agreement shall, where the reference is to two, arbitrators, be deemed to include a provision that the two arbitrators may appoint an umpire at any time after they are themselves appointed and shall do so forthwith if they cannot agree.

(2) Unless a contrary intention is expressed therein, every arbitration agreement shall, where such a provision is applicable to the reference, be deemed to include a provision that if the arbitrators have delivered to any party to the arbitration agreement, or to the umpire, a notice in writing stating that they cannot agree, the umpire may forthwith enter on the reference in lieu of the arbitrators.

(3) At any time after the appointment of an umpire, however appointed, the High Court may, on the application of any party to the reference and notwithstanding anything to the contrary in the arbitration agreement, order that the umpire shall enter upon the reference in lieu of the arbitrators and as if he were a sole arbitrator.

[Majority award of three arbitrators
9. Unless the contrary intention is expressed in the arbitration agreement, in any case where there is a reference to three arbitrators, the award of any two of the arbitrators shall be binding.]

[Substituted by Arbitration Act 1979, s.6(2).]

Power of court in certain cases to appoint an arbitrator or umpire
[As amended by Arbitration Act 1979, s.6(3/4)]

10. In any of the following cases –

(*a*) where an arbitration agreement provides that the reference shall be to a single arbitrator, and all the parties do not, after differences have arisen, concur in the appointment of an arbitrator;

(*b*) if an appointed arbitrator refuses to act, or is incapable of acting or dies, and the arbitration agreement does not show that it was intended that the vacancy should not be supplied and the parties do not supply the vacancy;

(*c*) where the parties or two arbitrators are required or are at liberty to appoint an umpire or third arbitrator and do not appoint him;

(*d*) where an appointed umpire or third arbitrator refuses to act, or is incapable of acting, or dies, and the arbitration agreement does not show that it was intended that the vacancy should not be supplied, and the parties or arbitrators do not supply the vacancy;

any party may serve the other parties or the arbitrators, as the case may be, with a written notice to appoint or, as the case may be, concur in appointing, an arbitrator, umpire or third arbitrator, and if the appointment is not made within seven clear days after the service of the notice, the High Court or a judge thereof may, on application by the party who gave the notice, appoint an arbitrator, umpire or third arbitrator who shall have the like powers to act in the reference and make an award as if he had been appointed by consent of all parties.

(2) In any case where –

(*a*) an arbitration agreement provides for the appointment of an arbitrator or umpire by a person who is neither one of the parties nor an existing arbitrator (whether the provision applies directly or in default of agreement by the parties or otherwise), and

(*b*) that person refuses to make the appointment or does not make it within the time specified in the agreement or, if no time is so specified, within a reasonable time,

any party to the agreement may serve the person in question with a written notice to appoint an arbitrator or umpire and, if the appointment is not made within seven clear days after the service of the notice, the High Court or a judge thereof may, on the application of the party who gave the notice, appoint an arbitrator or umpire who shall have the like powers to act in the reference and make an award as if he had been appointed in accordance with the terms of the agreement.

Reference to official referee

11. Where an arbitration agreement provides that the reference shall be to an official referee, any official referee to whom application is made shall, subject to any order of the High Court or a judge thereof as to transfer or otherwise, hear and determine the matters agreed to be referred.

Conduct of Proceedings, Witnesses, etc.

Conduct of proceedings, witnesses, etc.

12. – (1) Unless a contrary intention is expressed therein, every arbitration agreement shall, where such a provision is applicable to the reference, be deemed to contain a provision that the parties to the reference, and all persons claiming through them respectively, shall, subject to any legal objection, submit to be examined by the arbitrator or umpire, on oath or affirmation, in relation to the matters in dispute, and shall, subject as aforesaid, produce before the arbitrator or umpire all documents within their possession or power respectively which may be required or called for, and do all other things which during the proceedings on the reference the arbitrator or umpire may require.

(2) Unless a contrary intention is expressed therein, every arbitration agreement shall, where such a provision is applicable to the reference, be deemed to contain a provision that the witnesses on the reference shall, if the arbitrator or umpire thinks fit, be examined on oath or affirmation.

(3) An arbitrator or umpire shall, unless a contrary intention is expressed in the abitration agreement, have power to administer oaths to, or take the affirmations of, the parties to and witnesses on a reference under the agreement.

(4) Any party to a reference under an arbitration agreement may sue out a writ of subpoena ad testificandum or a writ of subpoena duces tecum, but no person shall be compelled under any such writ to produce any document which he could not be compelled to produce on the trial of an action, and the High Court or a judge thereof may order that a writ of subpoena ad testificandum or of subpoena duces tecum shall issue to compel the attendance before an arbitrator or umpire of a witness wherever he may be within the United Kingdom.

(5) The High Court or a judge thereof may also order that a writ of habeas corpus ad testificandum shall issue to bring up a prisoner for examination before an arbitrator or umpire.

(6) The High Court shall have, for the purpose of and in relation to a reference, the same power of making orders in respect of –

(*a*) security for costs;

(*b*) discovery of documents and interrogatories;

(*c*) the giving of evidence by affidavit;

(*d*) examination on oath of any witness before an officer of the High Court or any other person, and the issue of a commission or request for the examination of a witness out of the jurisdiction;

(*e*) the preservation, interim custody or sale of any goods which are the subject matter of the reference;

(*f*) securing the amount in dispute in the reference;

(*g*) the detention, preservation or inspection of any property or thing which is the subject of the reference or as to which any question may arise therein, and authorising for any of the purposes aforesaid any persons to enter upon or into any land or building in the possession of any party to the reference, or authorising any samples to be taken or any observation to be made or experiment to be tried which may be necessary or expedient for the purpose of obtaining full information or evidence; and

(*h*) interim injunctions or the appointment of a receiver;

as it has for the purpose of and in relation to an action or matter in the High Court:

Provided that nothing in this subsection shall be taken to prejudice any power which may be vested in an arbitrator or umpire of making orders with respect to any of the matters aforesaid.

Provisions as to Awards

Time for making award

13. – (1) Subject to the provisions of subsection (2) of section twenty-two of this Act, and anything to the contrary in the arbitration agreement, an arbitrator or umpire shall have power to make an award at any time.

(2) The time, if any, limited for making an award, whether under this Act or otherwise, may from time to time be enlarged by order of the High Court or a judge thereof, whether that time has expired or not.

(3) The High Court may, on the application of any party to a reference, remove an arbitrator or umpire who fails to use all reasonable dispatch in

entering on and proceeding with the reference and making an award, and an arbitrator or umpire who is removed by the High Court under this subsection shall not be entitled to receive any remuneration in respect of his services.

For the purposes of this subsection, the expression 'proceeding with a reference' includes, in a case where two arbitrators are unable to agree, giving notice of the fact to the parties and to the umpire.

Interim awards
14. Unless a contrary intention is expressed therein, every arbitration agreement shall, where such a provision is applicable to the reference, be deemed to contain a provision that the arbitrator or umpire may, if he thinks fit, make an interim award, and any reference in this Part of the Act to an award includes a reference to an interim award.

Specific performance
15. Unless a contrary intention is expressed therein, every arbitration agreement shall, where such a provision is applicable to the reference, be deemed to contain a provision that the arbitrator or umpire shall have the same power as the High Court to order specific performance of any contract other than a contract relating to land or any interest in land.

Awards to be final
16. Unless a contrary intention is expressed therein, every arbitration agreement shall, where such a provision is applicable to the reference, be deemed to contain a provision that the award to be made by the arbitrator or umpire shall be final and binding on the parties and the persons claiming under them respectively.

Power to correct slips
17. Unless a contrary intention is expressed in the arbitration agreement, the arbitrator or umpire shall have power to correct in an award any clerical mistake or error arising from any accidental slip or omission.

Costs, Fees and Interest

Costs
18. – (1) Unless a contrary intention is expressed therein, every arbitration agreement shall be deemed to include a provision that the costs of the reference and award shall be in the discretion of the arbitrator or umpire, who may direct to and by whom and in what manner those costs or any part thereof shall be paid, and may tax or settle the amount of costs to be so paid or any part thereof, and may award costs to be paid as between solicitor and client.

(2) Any costs directed by an award to be paid shall, unless the award otherwise directs, be taxable in the High Court.

(3) Any provision in an arbitration agreement to the effect that the parties or any party thereto shall in any event pay their or his own costs of the reference or award or any part thereof shall be void, and this Part of this Act, shall in the case of an arbitration agreement containing any such provision, have effect as if that provision were not contained therein:

Provided that nothing in this subsection shall invalidate such a provision when it is a part of an agreement to submit to arbitration a dispute which has arisen before the making of that agreement.

(4) If no provision is made by an award with respect to the costs of the reference, any party to the reference may, within fourteen days of the publication of the award or such further time as the High Court or a judge thereof may direct, apply to the arbitrator for an order directing by and to whom those costs shall be paid, and thereupon the arbitrator shall, after hearing any party who may desire to be heard, amend his award by adding thereto such directions as he may think proper with respect to the payment of the costs of the reference.

(5) Section sixty-nine of the Solicitors Act 1932 [see now section 73 of the Solicitors Act 1973] (which empowers a court before which any proceeding is being heard or is pending to charge property recovered or preserved in the proceeding with the payment of solicitors' costs) shall apply as if an arbitration were a proceeding in the High Court and the High Court may make declarations and orders accordingly.

Taxation of arbitrator's or umpire's fees

19. - (1) If in any case an arbitrator or umpire refuses to deliver his award except on payment of the fees demanded by him, the High Court may, on an application for the purpose, order that the arbitrator or umpire shall deliver the award to the applicant on payment into court by the applicant of the fees demanded, and further that the fees demanded shall be taxed by the taxing officer and that out of the money paid into court there shall be paid out to the arbitrator or umpire by way of fees such sums as may be found reasonable on taxation and that the balance of the money, if any, shall be paid out to the applicant.

(2) An application for the purpose of this section may be made by any party to the reference unless the fees demanded have been fixed by a written agreement between him and the arbitrator or umpire.

(3) A taxation of fees under this section may be reviewed in the same manner as a taxation of costs.

(4) The arbitrator or umpire shall be entitled to appear and be heard on any taxation or review of taxation under this section.

Power of arbitrator to award interest
19A. - (1) Unless a contrary intention is expressed therein, every arbitration agreement shall, where such a provision is applicable to the reference, be deemed to contain a provision that the arbitrator or umpire may, if he thinks fit, award simple interest at such rate as he thinks fit –

(a) on any sum which is the subject of the reference but which is paid before the award, for such period ending not later than the date of the payment as he thinks fit; and

(b) on any sum which he awards, for such period ending not later than the date of the award as he thinks fit.

(2) The power to award interest conferred on an arbitrator or umpire by subsection (1) above is without prejudice to any other power of an arbitrator or umpire to award interest.

[Added by Administration of Justice Act 1982, Section 15(6)]

Interest on awards
20. A sum directed to be paid by an award shall, unless the award otherwise directs, carry interest as from the date of the award and at the same rate as a judgment debt.

Special Cases, Remission and Setting aside of Awards, etc.

Statement of case
21. - *(1) An arbitrator or umpire may, and shall if so directed by the High Court, state –*

(a) any question of law arising in the course of the reference, or

(b) an award or any part of an award,

in the form of a special case for the decision of the High Court.

(2) A special case with respect to an interim award or with respect to a question of law arising in the course of a reference may be stated, or may be directed by the High Court to be stated, notwithstanding that proceedings under the reference are still pending.

(3) A decision of the High Court under this section shall be deemed to be a judgment of the Court within the meaning of section twenty-seven of the Supreme Court of Judicature (Consolidation) Act 1925 (which relates to the jurisdiction of the Court of Appeal to hear and determine appeals from and

judgment of the High Court), but no appeal shall lie from the decision of the High Court on any case stated under paragraph (a) of subsection (1) of this section without the leave of the High Court or of the Court of Appeal.

[Repealed by Arbitration Act 1979, s.8(3)(*b*)]

Power to remit award

22. – (1) In all cases of reference to arbitration the High Court or a judge thereof may from time to time remit the matters referred, or any of them, to the reconsideration of the arbitrator or umpire.

(2) Where an award is remitted, the arbitrator or umpire shall, unless the order otherwise directs, make his award within three months after the date of the order.

Removal of arbitrator and setting aside of award

23. – (1) Where an arbitrator or umpire has misconducted himself or the proceedings, the High Court may remove him.

(2) Where an arbitrator or umpire has misconducted himself or the proceedings, or an arbitration or award has been properly procured the High Court may set the award aside.

(3) Where an application is made to set aside an award, the High Court may order that any money made payable by the award shall be brought into court or otherwise secured pending the determination of the application.

Power of court to give relief where arbitrator is not impartial or the dispute involves question of fraud

24. – (1) Where an agreement between any parties provides that disputes which may arise in the future between them shall be referred to an arbitrator named or designated in the agreement, and after a dispute has arisen any party applies, on the ground that the arbitrator so named or designated is not or may not be impartial, for leave to revoke the authority of the arbitrator or for an injunction to restrain any other party or the arbitrator from proceeding with the arbitration, it shall not be a ground for refusing the application that the said party at the time when he made the agreement knew, or ought to have known, that the arbitrator, by reason of his relation towards any other party to the agreement or of his connection with the subject referred, might not be capable of impartiality.

(2) Where an agreement between any parties provides that disputes which may arise in the future between them shall be referred to arbitration, and a dispute which so arises involves the question whether any such party has been guilty of fraud, the High Court shall, so far as may be necessary to enable that question to be determined by the High Court, have power to

order that the agreement shall cease to have effect and power to give leave to revoke the authority of any arbitrator or umpire appointed by or by virtue of the agreement.

[See restriction in Arbitration Act 1979, s.3(3)]

(3) In any case where by virtue of this section the High Court has power to order that an arbitration agreement shall cease to have effect or to give leave to revoke the authority of an arbitrator or umpire the High Court may refuse to stay any action brought in breach of the agreement.

Power of court where arbitrator is removed or authority of arbitrator is revoked

25. – (1) Where an arbitrator (not being a sole arbitrator), or two or more arbitrators (not being all the arbitrators) or an umpire who has not entered on the reference is or are removed by the High Court or the Court of Appeal, the High Court or the Court of Appeal as the case may be, may, on the application of any party to the arbitration agreement, appoint a person or persons to act as arbitrator or arbitrators or umpire in place of the person or persons so removed.

(2) Where the authority of an arbitrator or arbitrators or umpire is revoked by leave of the High Court or the Court of Appeal, or a sole arbitrator or all the arbitrators or an umpire who has entered on the reference is or are removed by the High Court or the Court of Appeal, the High Court or the Court of Appeal as the case may be, may, on the application of any party to the arbitration agreement, either–

  (*a*) appoint a person to act as sole arbitrator in place of the person or persons removed; or

  (*b*) order that the arbitration agreement shall cease to have effect with respect to the dispute referred.

(3) A person appointed under this section by the High Court or the Court of Appeal as an arbitrator or umpire shall have the like power to act in the reference and to make an award as if he had been appointed in accordance with the terms of the arbitration agreement.

(4) Where it is provided (whether by means of a provision in the arbitration agreement or otherwise) that an award under an arbitration agreement shall be a condition precedent to the bringing of an action with respect to any matter to which the agreement applies, the High Court or the Court of Appeal, if it orders (whether under this section or under any other enactment) that the agreement shall cease to have effect as regards any particular dispute, may further order that the provision making an award a condition precedent to the bringing of an action shall also cease to have effect as regards that dispute.

Enforcement of Award

Enforcement of award
26. - (1) An award on an arbitration agreement may, by leave of the High Court or a judge thereof, be enforced in the same manner as a judgment or order to the same effect, and where leave is so given, judgment may be entered in terms of the award.

(2) If -

(*a*) the amount sought to be recovered does not exceed the current limit on jurisdiction in section 40 of the County Courts Act 1959, and

(*b*) a county court so orders,

it shall be recoverable (by execution issued from the county court or otherwise) as if payable under an order of that court and shall not be enforceable under subsection (1) above.

(3) An application to the High Court under this section shall preclude an application to a county court and an application to a county court under this section shall preclude an application to the High Court.

[As amended by Administration of Justice Act 1977, s.17(2)]

Miscellaneous

Power of court to extend time for commencing arbitration proceedings
27. Where the terms of an agreement to refer future disputes to arbitration provide that any claims to which the agreement applies shall be barred unless notice to appoint an arbitrator is given or an arbitrator is appointed or some other step to commence arbitration proceedings is taken within a time fixed by the agreement, and a dispute arises to which the agreement applies, the High Court, if it is of opinion that in the circumstances of the case undue hardship would otherwise be caused, and notwithstanding that the time so fixed has expired, may, on such terms, if any, as the justice of the case may require, but without prejudice to the provisions of any enactment limiting the time for the commencement of arbitration proceedings, extend the time for such period as it thinks proper.

Terms as to costs, etc.
28. Any order made under this Part of this Act may be made on such terms as to costs or otherwise as the authority making the order thinks just.

[As amended by Arbitration Act 1975, s.8(2)(*b*)]

Extension of section 496 of the Merchant Shipping Act 1894

29. - (1) In subsection (3) of section four hundred and ninety-six of the Merchant Shipping Act 1894 (which requires a sum deposited with a wharfinger by an owner of goods to be repaid unless legal proceedings are instituted by the shipowner), the expression 'legal proceedings' shall be deemed to include arbitration.

(2) For the purposes of the said section four hundred and ninety-six, as amended by this section, an arbitration shall be deemed to be commenced when one party to the arbitration agreement serves on the other party or parties a notice requiring him or them to appoint or concur in appointing an arbitrator, or, where the arbitration agreement provides that the reference shall be to a person named or designated in the agreement, requiring him or them to submit the dispute to the person so named or designated.

(3) Any such notice as is mentioned in subsection (2) of this section may be served either –

(*a*) by delivering it to the person on whom it is to be served; or

(*b*) by leaving it at the usual or last known place of abode in England of that person; or

(*c*) by sending it by post in a registered letter addressed to that person at his usual or last known place of abode in England;

as well as in any other manner provided in the arbitration agreement; and where a notice is sent by post in manner prescribed by paragraph (*c*) of this subsection, service thereof shall, unless the contrary is proved, be deemed to have been effected at the time at which the letter would have been delivered in the ordinary course of post.

Crown to be bound

30. This part of this Act [. . .] shall apply to any arbitration to which his Majesty, either in right of the Crown or of the Duchy of Lancaster or otherwise, or the Duke of Cornwall, is a party.

[As amended by Arbitration Act 1975, s.8(2)(*c*)]

Application of Part I to statutory arbitrations

31. - (1) Subject to the provisions of section thirty-three of this Act, this Part of this Act, expect the provisions thereof in subsection (2) of this section, shall apply to every arbitration under any other Act (whether passed before or after the commencement of this Act) as if the arbitration were pursuant to an arbitration agreement and as if that other Act were an arbitration agreement, except in so far as this Act is inconsistent with that other Act or with any rules or procedure authorised or recognised thereby.

(2) The provisions referred to in subsection (1) of this section are subsection (1) of section two, section three, section five, subsection (3) of section eighteen and sections twenty-four, twenty-five, twenty-seven and twenty-nine.

[See declaration in Arbitration Act 1979, s.7(3). As amended by Arbitration Act 1975, s.8(2)(*a*)]

Meaning of 'arbitration agreement'

32. In this Part of this Act, unless the context otherwise requires, the expression 'arbitration agreement' means a written agreement to submit present or future differences to arbitration, whether an arbitrator is named therein or not.

Operation of Part I

33. This Part of this Act shall not affect any arbitration commenced (within the meaning of subsection (2) of section twenty-nine of this Act) before the commencement of this Act, but shall apply to an arbitration so commenced after the commencement of this Act under an agreement made before the commencement of this Act.

Extent of Part I

34. None of the provisions of this Part of this Act shall extend to Scotland or Northern Ireland.

[As amended by Arbitration Act 1975, s.8(2)(*e*)]

Part II

Enforcement of Certain Foreign Awards

Awards to which Part II applies

35. – (1) This Part of this Act applies to any award made after the twenty-eighth day of July, nineteen hundred and twenty-four –

(*a*) in pursuance of an agreement for arbitration to which the protocol set out in the First Schedule to this Act applies; and

(*b*) between persons of whom one is subject to the jurisdiction of some one of such Powers as His Majesty, being satisfied that reciprocal provisions have been made, may by Order in Council declare to be parties to the convention set out in the Second Schedule to this Act, and of whom the other is subject to the jurisdiction of some other of the Powers aforesaid; and

(*c*) in one of such territories as His Majesty, being satisfied that reciprocal provisions have been made, may by Order in Council declare to be territories to which the said convention applies;

and an award to which this Part of this Act applies is in this Part of this Act referred to as 'a foreign award.'

(2) His Majesty may by a subsequent Order in Council vary or revoke any Order previously made under this section.

(3) Any Order in Council under section one of the Arbitration (Foreign Awards) Act 1930, which is in force at the commencement of this Act shall have effect as if it had been made under this section.

Effect of foreign awards
36. - (1) A foreign award shall, subject to the provisions of this Part of this Act, be enforceable in England either by action or in the same manner as the award of an arbitrator is enforceable by virtue of section twenty-six of this Act.

(2) Any foreign award which would be enforceable under this Part of this Act shall be treated as binding for all purpose on the persons as between whom it was made, and may accordingly be relied on by any of those persons by way of defence, set off or otherwise in any legal proceedings in England, and any references in this Part of this Act to enforcing a foreign award shall be construed as including references to relying on an award.

Conditions for enforcement of foreign awards
37. - (1) In order that a foreign award may be enforceable under this part of this Act it must have -

(*a*) been made in pursuance of an agreement for arbitration which was valid under the law by which it was governed;

(*b*) been made by the tribunal provided for in the agreement or constituted in manner agreed upon by the parties;

(*c*) been made in conformity with the law governing the arbitration procedure;

(*d*) become final in the country in which it was made;

(*e*) been in respect of a matter which may lawfully be referred to arbitration under the law of England;

and the enforcement thereof must not be contrary to the public policy or the law of England.

(2) Subject to the provisions of this subsection, a foreign award shall not be enforceable under this Part of this Act if the court dealing with the case is satisfied that -

(a) the award has been annulled in the country in which it was made; or

(b) the party against whom it is sought to enforce the award was not given notice of the arbitration proceedings in sufficient time to enable him to present his case, or was under some legal incapacity and was not properly represented; or

(c) the award does not deal with all the questions referred or contains decisions on matters beyond the scope of the agreement for arbitration.

Provided that, if the award does not deal with all the questions referred, the court may, if it thinks fit, either postpone the enforcement of the award or order its enforcement subject to the giving of such security by the person seeking to enforce it as the court may think fit.

(3) If a party seeking to resist the enforcement of a foreign award proves that there is any ground other than the non-existence of the conditions specified in paragraphs (a), (b), and (c) of subsection (1) of this section, or the existence of the conditions specified in paragraphs (b) and (c) of subsection (2) of this section, entitling him to contest the validity of the award, the court may, if it thinks fit, either refuse to enforce the award or adjourn the hearing until after the expiration of such period as appears to the court to be reasonably sufficient to enable that party to take the necessary steps to have the award annulled by the competent tribunal.

Evidence

38. – (1) The party seeking to enforce a foreign award must produce –

(a) the original award or a copy thereof duly authenticated in manner required by the law of the country in which it was made; and

(b) evidence proving that the award has become final; and

(c) such evidence as may be necessary to prove that the award is a foreign award and that the conditions mentioned in paragraphs (a), (b), and (c) of subsection (1) of the last foregoing section are satisfied.

(2) In any case where any document required to be produced under subsection (1) of this section is in a foreign language, it shall be the duty of the party seeking to enforce the award to produce a translation certified as correct by a diplomatic or consular agent of the country to which that party belongs, or certified as correct in such other manner as may be sufficient according to the law of England.

(3) Subject to the provisions of this section, rules of court may be made under section [84 of the Supreme Court Act 1981] with respect to the evidence which must be furnished by a party seeking to enforce an award under this Part of this Act.

[In relation to Northern Ireland see s.55 of the Judicature (Northern Ireland) Act 1978. The subs was substituted by s.152(1) of and Sch.5 to the Supreme Court Act, 1981]

Meaning of 'final award'
39. For the purposes of this Part of this Act, an award shall not be deemed final if any proceedings for the purpose of contesting the validity of the award are pending in the country in which it was made.

Saving for rights, etc.
40. Nothing in this Part of this Act shall–

   (*a*) prejudice any rights which any person would have had of enforcing in England any award or of availing himself in England of any award if neither this Part of this Act nor Part I of the Arbitration (Foreign Awards) Act 1930 had been enacted; or

   (*b*) apply to any award made on an arbitration agreement governed by the law of England.

Application of Part II to Scotland
41. – (1) The following provisions of this section shall have effect for the purpose of the application of this Part of this Act to Scotland.

(2) For the references to England there shall be substituted references to Scotland.

(3) For subsection (1) of section thirty-six there shall be substituted the following subsection:

   '(1) A foreign award shall, subject to the provisions of this Part of this Act, be enforceable by action, or, if the agreement for arbitration contains consent to the registration of the award in the Books of Council and Session for execution and the award is so registered, it shall, subject as aforesaid, be enforceable by summary diligence.'

(4) For subsections (3) of section thirty-eight there shall be substituted the following subsection:

   '(3) The Court of Session shall, subject to the provisions of this section, have power to make provision by Act of Sederunt with respect to the evidence which must be furnished by a party seeking to enforce in Scotland an award under this Part of this Act.'

[As amended by Law Reform (Miscellaneous Provisions) (Scotland) Act 1966 (c.19)]

Application of Part II to Northern Ireland
42. – (1) The following provisions of this section shall have effect for the purpose of the application of this Part of this Act to Northern Ireland.

(2) For the references to England there shall be substituted references to Northern Ireland.

(3) For subsection (1) of section thirty-six there shall be substituted the following subsection:—

'(1) A foreign award shall, subject to the provisions of this Part of this Act, be enforceable either by action or in the same manner as the award of an arbitrator under the provisions of the Common Law Procedure Amendment Act (Ireland) 1856 was enforceable at the date of the passing of the Arbitration (Foreign Awards) Act 1930.'

(4) [*Repealed by Judicature (Northern Ireland) Act 1978, Sched. 7.*]

Saving for pending proceedings
43. [*Repealed by the Statute Law (Repeals) Act 1978, Sched. 1.*]

PART III

GENERAL

Short title, commencement and repeal
44. - (1) This Act may be cited as the Arbitration Act 1950.

(2) This Act shall come into operation on the first day of September, nineteen hundred and fifty.

(3) The Arbitration Act 1889, the Arbitration Clauses (Protocol) Act 1924 and the Arbitration Act 1934 are hereby repealed except in relation to arbitrations commenced (within the meaning of subsection (2) of section twenty-nine of this Act) before the commencement of this Act, and the Arbitration (Foreign Awards) Act 1930 is hereby repealed; and any reference in any Act or other document to any enactment hereby repealed shall be construed as including a reference to the corresponding provision of this Act.

SCHEDULES

Sections 4, 35.

FIRST SCHEDULE

PROTOCOL ON ARBITRATION CLAUSES SIGNED ON BEHALF OF HIS MAJESTY AT A MEETING OF THE ASSEMBLY OF THE LEAGUE OF NATIONS HELD ON THE TWENTY-FOURTH DAY OF SEPTEMBER, NINETEEN HUNDRED AND TWENTY-THREE

The undersigned, being duly authorised, declare that they accept, on behalf of the countries which they represent, the following provisions; –

1. Each of the Contracting States recognises the validity of an agreement whether relating to existing or future differences between parties, subject respectively to the jurisdiction of different Contracting States by which the parties to a contract agree to submit to arbitration all or any differences that may arise in connection with such contract relating to commercial matters or to any other matter capable of settlement by arbitration, whether or not the arbitration is to take place in a country to whose jurisdiction none of the parties is subject. Each Contracting State reserves the right to limit the obligation mentioned above to contracts which are considered as commercial under its national law. Any Contracting State which avails itself of this right will notify the Secretary-General of the League of Nations, in order that the other Contracting States may be so informed.

2. The arbitral procedure, including the constitution of the arbitral tribunal, shall be governed by the will of the parties and by the law of the country in whose territory the arbitration takes place.

The Contracting States agree to facilitate all steps in the procedure which require to be taken in their own territories, in accordance with the provisions of their law governing arbitral procedure applicable to existing differences.

3. Each Contracting State undertakes to ensure the execution by its authorities and in accordance with the provisions of its national laws of arbitral awards made in its own territory under the preceding articles.

4. The tribunals of the Contracting Parties, on being seized of a dispute regarding a contract made between persons to whom Article 1 applies and including an arbitration agreement whether referring to present or future differences which is valid in virtue of the said article and capable of being carried into effect, shall refer the parties on the application of either of them to the decision of the arbitrators.

Such reference shall not prejudice the competence of the judicial tribunals in case the agreement of the arbitration cannot proceed or become inoperative.

5. The present Protocol, which shall remain open for signature by all States, shall be ratified. The ratifications shall be deposited as soon as possible with the Secretary-General of the League of Nations, who shall notify such deposit to all the signatory States.

6. The present Protocol shall come into force as soon as two ratifications have been deposited. Thereafter it will take effect, in the case of each Contracting State, one month after the notification by the Secretary-General of the deposit of its ratification.

7. The present Protocol may be denounced by any Contracting State on giving one year's notice. Denunciation shall be effected by a notification addressed to the Secretary-General of the League, who will immediately

transmit copies of such notification to all the other signatory States and inform them on the date of which it was received. The denunciation shall take effect one year after the date on which it was notified to the Secretary-General and shall operate only in respect of the notifying State.

8. The Contracting States may declare that their acceptance of the present Protocol does not include any or all of the under-mentioned territories: that is to say, their colonies, overseas possessions or territories, protectorates or the territories over which they exercise a mandate.

The said States may subsequently adhere separately on behalf of any territory thus excluded. The Secretary-General of the League of Nations shall be informed as soon as possible of such adhesions. He shall notify such adhesions to all signatory States. They will take effect one month after the notification by the Secretary-General.

The Contracting States may also denounce the Protocol separately on behalf of any of the territories referred to above. Article 7 applies to such denunication.

This Protocol formed the Schedule to the Arbitration Clauses (Protocol) Act 1924.

Section 35.

SECOND SCHEDULE

Convention on the Execution of Foreign Arbitral Awards signed at Geneva on behalf of His Majesty on the twenty-sixth day of September, nineteen hundred and twenty-seven

Article 1

In the territories of any High Contracting Party to which the present Convention applies, an arbitral award made in pursuance of an agreement, whether relating to existing or future differences (hereinafter called 'a submission to arbitration') covered by the Protocol on Arbitration Clauses, opened at Geneva on September 24, 1923, shall be recognised as binding and shall be enforced in accordance with the rules of the procedure of the territory where the award is relied upon, providing that the said award has been made in a territory of one of the High Contracting Parties to which the present Convention applies and between persons who are subject to the jurisdiction of one of the High Contracting Parties.

To obtain such recognition or enforcement, it shall, further, be necessary –

(*a*) that the award has been made in pursuance of a submission to arbitration which is valid under the law applicable thereto;

(b) That the subject-matter of the award is capable of settlement by arbitration under the law of the country in which the award is sought to be relied upon;

(c) That the award has been made by the Arbitral Tribunal provided for in the submission to arbitration or constituted in the manner agreed upon by the parties and in conformity with the law governing the arbitration procedure;

(d) That the award has become final in the country in which it has been made, in the sense that it will not be considered as such if it is open to *opposition, appeul* or *pourvoi en cessation* (in the countries where such forms of procedure exist) or if it is proved that any proceedings for the purpose of contesting the validity of the award are pending;

(e) That the recognition or enforcement of the award is not contrary to the public policy or to the principles of the law of the country in which it is sought to be relied upon.

Article 2

Even if the conditions laid down in Article 1 hereof are fulfilled, recognition and enforcement of the award shall be refused if the Court is satisfied –

(a) That the award has been annulled in the country in which it was made;

(b) That the party against whom it is sought to use the award was not given notice of the arbitration proceedings in sufficient time to enable him to present his case; or that, being under a legal incapacity, he was not properly represented;

(c) That the award does not deal with the differences contemplated by or falling within the terms of the submission to arbitration or that it contains decisions on matters beyond the scope of the submission to arbitration.

If the award has not covered all the questions submitted to the arbitral tribunal, the competent authority of the country where recognition or enforcement of the award is sought can, if it thinks fit, postpone such recognition or enforcement or grant is subject to such guarantee as that authority may decide.

Article 3

If the party against whom the award has been made proves that, under the law governing the arbitration procedure, there is a ground, other than the grounds referred to in Article 1(a) and (c), and Article 2(b) and (c) entitling him to contest the validity of the award in a Court of Law, the Court may, if it thinks fit, either refuse recognition or enforcement of the award or adjourn the consideration thereof, giving such party a reasonable time within which to have the award annulled by the competent tribunal.

ARTICLE 4

The party relying upon an award or claiming its enforcement must supply, in particular –
  (1) The original award or a copy thereof duly authenticated, according to the requirements of the law of the country in which it was made;
  (2) Documentary or other evidence to prove that the award has become final, in the sense defined in Article 1(*d*), in the country in which it was made;
  (3) When necessary, documentary or other evidence to prove that the conditions laid down in Article 1, paragraph 1 and paragraph 2(*a*) and (*c*), have been fulfilled.

A translation of the award and of the other documents mentioned in this Article into the official language of the country where the award is sought to be relied upon may be demanded. Such translation must be certified correct by a diplomatic or consular agent of the country to which the party who seeks to rely upon the award belongs or by a sworn translator of the country where the award is sought to be relied upon.

ARTICLE 5

The provisions of the above Articles shall not deprive any interested party of the right of availing himself of an arbitral award in the manner and to the extent allowed by the law or the treaties of the country where such award is sought to be relied upon.

ARTICLE 6

The present Convention applies only to arbitral awards made after the coming into force of the Protocol on Arbitration Clauses, opened at Geneva on September 24, 1923.

Article 7

The present Convention, which will remain open to the signature of all the signatories of the Protocol of 1923 on Arbitration Clauses, shall be ratified.

It may be ratified only on behalf of those Members of the League of Nations and non-Member States on whose behalf the Protocol of 1923 shall have been ratified.

Ratifications shall be deposited as soon as possible with the Secretary-General of the League of Nations, who will notify such deposit to all the signatories.

Article 8

The present Convention shall come into force three months after it shall have been ratified on behalf of two High Contracting Parties. Thereafter, it shall take effect, in the case of each High Contracting Party, three months after the deposit of the ratification on its behalf with the Secretary-General of the League of Nations.

Article 9

The present Convention may be denounced on behalf of any Member of the League or non-Member State. Denunciation shall be notified in writing to the Secretary-General of the League of Nations, who will immediately send a copy thereof, certified to be in conformity with the notification, to all the other Contracting Parties, at the same time informing them of the date on which he received it.

The denunciation shall come into force only in respect of the High Contracting Party which shall have notified it and one year after such notification shall have reached the Secretary-General of the League of Nations.

The denunciation of the Protocol on Arbitration Clauses shall entail, ipso facto, the denunciation of the present Convention.

Article 10

The present Convention does not apply to the Colonies, Protectorates or territories under suzerainty or mandate of any High Contracting Party unless they are specially mentioned.

The application of this Convention to one or more of such Colonies, Protectorates or territories to which the Protocol on Arbitration Clauses, opened at Geneva on September 24th, 1923, applies, can be effected at any time by means of a declaration addressed to the Secretary-General of the League of Nations by one of the High Contracting Parties.

Such declaration shall take effect three months after the deposit thereof.

The High Contracting Parties can at any time denounce the Convention for all or any of the Colonies, Protectorates or territories referred to above. Article 9 hereof applies to such denunciation.

ARTICLE 11

A certified copy of the Present Convention shall be transmitted by the Secretary-General of the League of Nations to every Member of the League of Nations and to every non-Member which signs the same.

ARBITRATION ACT 1975

(1975, c. 3)

An Act to give effect to the New York Convention on the Recognition and enforcement of Foreign Arbitral Awards [25th Feb 1975]

Effect of Arbitration Agreement on Court Proceedings

Staying court proceedings where party proves arbitration agreement

1. – (1) If any party to an arbitration agreement to which this section applies, or any person claiming through or under him, commences any legal proceedings in any court against any other party to the agreement, or any person claiming through or under him, in respect of any matter agreed to be referred, any party to the proceedings may at any time after appearance, and before delivering any pleadings or taking any other steps in the proceedings, apply to the court to stay the proceedings; and the court, unless satisfied that the arbitration agreement is null and void, inoperative or incapable of being performed or that there is not in fact any dispute between the parties with regard to the matter agreed to be referred, shall make an order staying the proceedings.

(2) This section applies to any arbitration agreement which is not a domestic arbitration agreement; and neither section 4(1) of the Arbitration Act 1950 nor section 4 of the Arbitration Act (Northern Ireland) 1937 shall apply to an arbitration agreement to which this section applies.

(3) In the application of this section to Scotland, for the references to staying proceedings there shall be substituted references to sisting proceedings.

(4) In this section 'domestic arbitration agreement' means an arbitration agreement which does not provide, expressly or by implication, for arbitration in a State other than the United Kingdom and to which neither –

(a) an individual who is a national of, or habitually resident in, any State other than the United Kingdom; nor

(b) a body corporate which is incorporated in, or whose central management and control is exercised in, any State other than the United Kingdom;

is a party at the time the proceedings are commenced.

Enforcement of Convention Awards

Replacement of former provisions
2. Sections 3 to 6 of this Act shall have effect with respect to the enforcement of Convention awards; and where a Convention award would, but for this section, be also a foreign award within the meaning of Part II of the Arbitration Act 1950, that Part shall not apply to it.

Effect of Convention awards
3. – (1) A Convention award shall, subject to the following provisions of this Act, be enforceable –

(a) in England and Wales, either by action or in the same manner as the award of an arbitrator is enforceable by virtue of section 26 of the Arbitration Act 1950;

(b) in Scotland, either by action or, in a case where the arbitration agreement contains consent to the registration of the award in the Books of Council and Session for execution and the award is so registered, by summary diligence;

(c) in Northern Ireland, either by action or in the same manner as the award of an arbitrator is enforceable by virtue of section 16 of the Arbitration Act (Northern Ireland) 1937.

(2) Any Convention award which would be enforceable under this Act shall be treated as binding for all purposes on the persons as between whom it was made, and may accordingly be relied on by any of those persons by way of defence, set off or otherwise in any legal proceedings in the United Kingdom; and any reference in this Act to enforcing a Convention award shall be construed as including references to relying on such an award.

Evidence

4. The party seeking to enforce a Convention award must produce –

(a) the duly authenticated original award or a duly certified copy of it; and

(b) the original arbitration agreement or duly certified copy of it; and

(c) where the award or agreement is in a foreign language, a translation of it certified by an official or sworn translator or by a diplomatic or consular agent.

Refusal of enforcement

5. – (1) Enforcement of a Convention award shall not be refused except in the cases mentioned in this section.

(2) Enforcement of a Convention award may be refused if the person against whom it is invoked proves –

(a) that a party to the arbitration agreement was (under the law applicable to him) under some incapacity; or

(b) that the arbitration agreement was not valid under the law to which the parties subjected it or, failing any indication thereon, under the law of the country where the award was made; or

(c) that he was not given proper notice of the appointment of the arbitrator or of the arbitration proceedings or was otherwise unable to present his case; or

(d) (subject to subsection (4) of this section) that the award deals with a difference not contemplated by or not falling within the terms of the submission to arbitration or contains decisions on matters beyond the scope of the submission to arbitration; or

(e) that the composition of the arbitral authority or the arbitral procedure was not in accordance with the agreement of the parties or, failing such agreement, with the law of the country where the arbitration took place; or

(f) that the award has not yet become binding on the parties, or has been set aside or suspended by a competent authority of the country in which, or under the law of which, it was made.

(3) Enforcement of Convention award may also be refused if the award is in respect of a matter which is not capable of settlement by arbitration, or if it would be contrary to public policy to enforce the award.

(4) A Convention award which contains decisions on matters not submitted to arbitration may be enforced to the extent that it contains decisions on matters submitted to arbitration which can be separated from those on matters not so submitted.

(5) Where an application for the setting aside or suspension of a Convention award has been made to such a competent authority as is mentioned in subsection (2)(*f*) of this section, the court before which enforcement of the award is sought may, if it thinks fit, adjourn the proceedings and may, on the application of the party seeking to enforce the award, order the other party to give security.

Saving
6. Nothing in this Act shall prejudice any right to enforce or rely on an award otherwise than under this Act or Part II of the Arbitration Act 1950.

General

Interpretation
7. – (1) In this Act –

'arbitration agreement' means an agreement in writing (including an agreement contained in any exchange of letters or telegrams) to submit to arbitration present or future differences capable of settlement by arbitration;

'Convention award' means an award made in pursuance of an arbitration agreement in the territory of a State, other than the United Kingdom, which is a party to the New York Convention; and

'the New York Convention' means the Convention on the Recognition and Enforcement of Foreign Arbitral Awards adopted by the United Nations Conference on International Commercial Arbitration on 10th June 1958.

(2) If Her Majesty by Order in Council declares that any State specified in the Order is a party to the New York Convention the Order shall, while in force, be conclusive evidence that that State is a party to that Convention.

(3) An Order in Council under this section may be varied or revoked by a subsequent Order in Council.

Short title, repeals, commencement and extent
8. – (1) This Act may be cited as the Arbitration Act 1975.

(2) The following provisions of the Arbitration Act 1950 are hereby repealed, that is to say—

(*a*) section 4(2);

(*b*) in section 28 the proviso;

(*c*) in section 30 the words '(except the provisions of subsection (2) of section 4 thereof)';

(*d*) in section 31(2) the words 'subsection (2) of section 4'; and

(*e*) in section 34 the words from the beginning to 'save as aforesaid.'

(3) This Act shall come into operation on such date as the Secretary of State may by order made by statutory instrument appoint.

(4) This Act extends to Northern Ireland.

ARBITRATION ACT 1979

(1979, c. 42)

Judicial review of arbitration awards

1. - (1) In the Arbitration Act 1950 (in this Act referred to as 'the principal Act') section 21 (statement of case for a decision of the High Court) shall cease to have effect and, without prejudice to the right of appeal conferred by subsection (2) below, the High Court shall not have jurisdiction to set aside or remit an award on an arbitration agreement on the ground of errors of fact or law on the face of the award.

(2) Subject to subsection (3) below, an appeal shall lie to the High Court on any question of law arising out of an award made on an arbitration agreement; and on the determination of such an appeal the High Court may by order—

(*a*) confirm, vary or set aside the award; or

(*b*) remit the award to the reconsideration of the arbitrator or umpire together with the court's opinion on the question of law which was the subject of the appeal;

and where the award is remitted under paragraph (*b*) above the arbitrator or umpire shall, unless the order otherwise directs, make his award within three months after the date of the order.

(3) An appeal under this section may be brought by any of the parties to the reference—

(*a*) with the consent of all the other parties to the reference; or

(*b*) subject to section 3 below, with the leave of the court.

(4) The High Court shall not grant leave under subsection (3)(*b*) above unless it considers that, having regard to all the circumstances, the determination of the question of law concerned could substantially affect the rights of one or more of the parties to the arbitration agreement; and the court may make any leave which it gives conditional upon the applicant complying with such conditions as it considers appropriate.

(5) Subject to subsection (6) below, if an award is made and, on an application made by any of the parties to the reference, –

(*a*) with the consent of all the other parties to the reference, or

(*b*) subject to section 3 below, with the leave of the court,

it appears to the High Court that the award does not or does not sufficiently set out the reasons for the award, the court may order the arbitrator or umpire concerned to state the reasons for his award in sufficient detail to enable the court, should an appeal be brought under this section, to consider any question of law arising out of the award.

(6) In any case where an award is made without any reason being given, the High Court shall not make an order under subsection (5) above unless it is satisfied –

(*a*) that before the award was made one of the parties to the reference gave notice to the arbitrator or umpire concerned that a reasoned award would be required; or

(*b*) that there is some special reason why such a notice was not given.

[(6A) Unless the High Court gives leave, no appeal shall lie to the Court of Appeal from a decision of the High Court –

(*a*) to grant or refuse leave under subsection (3)(*b*) or (5)(*b*) above; or

(*b*) to make or not to make an order under subsection (5) above.]

(7) No appeal shall lie to the Court of Appeal from a decision of the High Court on an appeal under this section unless –

(*a*) the High Court or the Court of Appeal gives leave; and

(*b*) it is certified by the High Court that the question of law to which its decision relates either is one of general public importance or is one which for some other special reason should be considered by the Court of Appeal.

(8) Where the award of an arbiter or umpire is varied on appeal, the award as varied shall have effect (except for the purposes of this section) as if it were the award of the arbitrator or umpire.

Subs. (6A) added by the Supreme Court Act 1981, s. 148(1). This amendment does not apply to decisions of the High Court pronounced before January 1, 1982.

Determination of preliminary point of law by court
2. - (1) Subject to subsection (2) and section (3) below, on an application to the High Court made by any of the parties to a reference –

(*a*) with the consent of an arbitrator who has entered on the reference or, if an umpire has entered on the reference, with his consent, or

(*b*) with the consent of all the other parties

the High Court shall have jurisdiction to determine any question of law arising in the course of the reference.

(2) The High Court shall not entertain an application under subsection (1)(*a*) above with respect to any question of law unless it is satisfied that—

(*a*) the determination of the application might produce substantial savings in costs to the parties; and

(*b*) the question of law is one in respect of which leave to appeal would be likely to be given under section 1(3)(*b*) above.

[(2A) Unless the High Court gives leave, no appeal shall lie to the Court of Appeal from a decision of the High Court to entertain or not to entertain an application under subsection (1)(*a*) above.]

[Added by Supreme Court Act 1981 s. 148(3)]

(3) A decision of the High Court under [subsection (1) above] shall be deemed to be a judgment of the court within the meaning of section [16 of the Supreme Court Act 1981] (appeals to the Court of Appeal), but no appeal shall lie from such a decision unless –

(*a*) the High Court or the Court of Appeal gives leave; and

(*b*) it is certified by the High Court that the question of law to which its decision relates either is one of general public importance or is one which for some other special reason should be considered by the Court of Appeal.

[As amended by *ibid.*, s. 152(1) and Sched. 5.]

Exclusion agreements affecting rights under sections (1) and (2)

3. – (1) Subject to the following provisions of this section and section 4 below –

(*a*) the High Court shall not, under section 1(3)(*b*) above, grant leave to appeal with respect to a question of law arising out of an award, and

(*b*) the High Court shall not, under section 1(5)(*b*) above, grant leave to make an application with respect to an award, and

(*c*) no application may be made under section 2(1)(*a*) above with respect to a question of law,

if the parties to the reference in question have entered into an agreement in writing (in this section referred to as an 'exclusion agreement') which excludes the right of appeal under section 1 above in relation to that award or, in a case falling within paragraph (*c*) above, in relation to an award to which the determination of the question of law is material.

(2) An exclusion agreement may be expressed so as to relate to a particular award, to awards under a particular reference or to any other description of awards, whether arising out of the same reference or not; and an agreement may be an exclusion agreement for the purposes of this section whether it is entered into before or after the passing of this Act and whether or not it forms part of an arbitration agreement.

(3) In any case where –

(*a*) an arbitration agreement, other than a domestic arbitration agreement, provides for disputes between the parties to be referred to arbitration, and

(*b*) a dispute to which the agreement relates involves the question whether a party has been guilty of fraud, and

(*c*) the parties have entered into an exclusion agreement which is applicable to any award made on the reference of that dispute,

then, except in so far as the exclusion agreement otherwise provides, the High Court shall not exercise its powers under section 24(2) of the principal Act (to take steps necessary to enable the question to be determined by the High Court) in relation to that dispute.

(4) Except as provided by subsection (1) above, sections 1 and 2 above shall have effect notwithstanding anything in any agreement purporting –

(*a*) to prohibit or restrict access to the High Court; or

(*b*) to restrict the jurisdiction of that court; or

(*c*) to prohibit or restrict the making of a reasoned award.

(5) An exclusion agreement shall be of no effect in relation to an award made on, or a question of law arising in the course of a reference under, a statutory arbitration, that is to say, such an arbitration as is referred to in subsection (1) of section 31 of the principal Act.

(6) An exclusion agreement shall be of no effect in relation to an award made on, or a question of law arising in the course of a reference under, an arbitration agreement which is a domestic arbitration agreement unless the exclusion agreement is entered into after the commencement of the arbitration in which the award is made or, as the case may be, in which the question of law arises.

(7) In this section 'domestic arbitration agreement' means an arbitration agreement which does not provide, expressly or by implication, for arbitration in a State other than the United Kingdom and to which neither –

(*a*) an individual who is a national of, or habitually resident in, any State other than the United Kingdom, nor

(*b*) a body corporate which is incorporated in, or whose central management and control is exercised in, any State other than the United Kingdom,

is a party at the time the arbitration agreement is entered into.

Exclusion agreements not to apply in certain cases

4. – (1) Subject to subsection (3) below, if an arbitration award or a question of law arising in the course of a reference relates, in whole or in part to –

(*a*) a question or claim falling within the Admiralty jurisdiction of the High Court, or

(*b*) a dispute arising out of a contract of insurance, or

(*c*) a dispute arising out of a commodity contract,

an exclusion agreement shall have no effect in relation to the award or question unless either –

(i) the exclusion agreement is entered into after the commencement of the arbitration in which the award is made or, as the case may be, in which the question of law arises, or

(ii) the award or question relates to a contract which is expressed to be governed by a law other than the law of England and Wales.

(2) In subsection (1)(*c*) above 'commodity contract' means a contract –

(*a*) for the sale of goods regularly dealt with on a commodity market or exchange in England or Wales which is specified for the purposes of this section by an order made by the Secretary of State; and

(*b*) of a description so specified.

(3) The Secretary of State may by order provide that subsection (1) above –

(*a*) shall cease to have effect; or

(*b*) subject to such conditions as may be specified in the order, shall not apply to any exclusion agreement made in relation to an arbitration award of a description so specified;

and an order under this subsection may contain such supplementary, incidental and transitional provisions as appear to the Secretary of State to be necessary or expedient.

(4) The power to make an order under subsection (2) or subsection (3) above shall be exercisable by statutory instrument which shall be subject to annulment in pursuance of a resolution of either House of Parliament.

(5) In this section 'exclusion agreement' has the same meaning as in section 3 above.

Interlocutory orders
5. – (1) If any party to a reference under an arbitration agreement falls within the time specified in the order or, if no time is so specified, within a reasonable time to comply with an order made by the arbitrator or umpire in the course of the reference, then, on the application of the arbitrator or umpire or of any party to the reference, the High Court may make an order extending the powers of the arbitrator or umpire as mentioned in subsection (2) below.

(2) In an order made by the High Court under this section, the arbitrator or umpire shall have power, to the extent and subject to any conditions specified in that order, to continue with the reference in default of appearance or of any other act by one of the parties in like manner as a judge of the High Court might continue with proceedings in that court where a party fails to comply with an order of that court or a requirement of rules of court.

(3) Section 4(5) of the Administration of Justice Act 1970 (jurisdiction of the High Court to be exercisable by the Court of Appeal in relation to

judge-arbitrators and judge-umpires) shall not apply in relation to the power of the High Court to make an order under this section, but in the case of a reference to a judge-arbitrator or judge-umpire that power shall be exercisable as in the case of any other reference to arbitration and also by the judge-arbitrator or judge-umpire himself.

(4) Anything done by a judge-arbitrator or judge-umpire in the exercise of the power conferred by subsection (3) above shall be done by him in his capacity as judge of the High Court and have effect as if done by that court.

Minor amendments relating to awards and appointment of arbitrators and umpires

6. *These amendments to the 1950 Act have been incorporated in the Act as printed above on pp. 121ff.*

Application and interpretation of certain provisions of Part I of principal Act

7. - (1) References in the following provisions of Part I of the principal Act to that Part of that Act shall have effect as if the preceding provisions of this Act were included in that Part, namely, -

(*a*) section 14 (interim awards);

(*b*) section 28 (terms as to costs of orders);

(*c*) section 30 (Crown to be bound);

(*d*) section 31 (application to statutory arbitrations); and

(*e*) section 32 (meaning of 'arbitration agreement').

(2) Subsections (2) and (3) of section 29 of the principal Act shall apply to determine when an arbitration is deemed to be commenced for the purposes of this Act.

(3) For the avoidance of doubt, it is hereby declared that the reference in subsection (1) of section 31 of the principal Act (statutory arbitrations) to arbitration under any other Act does not extend to arbitration under [section 64 of the County Courts Act 1984] (cases in which proceedings are to be or may be referred to arbitration) and accordingly nothing in this Act or in Part I of the principal Act applies to arbitration under the said section 92.

Short title, commencement, repeals and extent

8. - (1) This Act may be cited as the Arbitration Act 1979.

(2) This Act shall come into operation on such day as the Secretary of State may appoint by order made by statutory instrument; and such an order -

(a) may appoint different days for different provisions of this Act and for the purposes of the operation of the same provision in relation to different descriptions of arbitration agreement; and

(b) may contain such supplementary, incidental and transitional provisions as appear to the Secretary of State to be necessary or expedient.

(3) In consequence of the preceding provisions of this Act, the following provisions are hereby repealed, namely –

(a) in paragraph (c) of section 10 of the principal Act the words from 'or where' to the end of the paragraph;

(b) section 21 of the principal Act;

(c) in paragraph 9 of Schedule 3 to the Administration of Justice Act 1970, in sub-paragraph (1) the words '21(1) and (2)' and sub-paragraph (2).

(4) This Act forms part of the law of England and Wales only.

CIVIL EVIDENCE ACT 1968

(1968, c. 64)

Part I

Hearsay Evidence

Hearsay evidence to be admissible only by virtue of this Act and other statutory provisions, or by agreement

1. – (1) In any civil proceedings a statement other than one made by a person while giving oral evidence in those proceedings shall be admissible as evidence of any fact stated therein to the extent that it is so admissible by virtue of any provision of this Part of this Act or by virtue of any other statutory provision or by agreement of the parties, but not otherwise.

(2) In this section 'statutory provision' means any provision contained in, or in an instrument made under, this or any other Act, including any Act passed after this Act.

Admissibility of out-of-court statements as evidence of acts stated

2. – (1) In any civil proceedings a statement made, whether orally or in a document or otherwise, by any person, whether called as a witness in those proceedings or not, shall, subject to this section and to rules of court, be admissible as evidence of any fact stated therein of which direct oral evidence by him would be admissible.

(2) Where in any civil proceedings a party desiring to give a statement in evidence by virtue of this section has called or intends to call as a witness in the proceedings the person by whom the statement was made, the statement –

(*a*) shall not be given in evidence by virtue of this section on behalf of that party without the leave of the court; and

(*b*) without prejudice to paragraph (*a*) above, shall not be given in evidence by virtue of this section on behalf of that party before the conclusion of the examination-in-chief of the person by whom it was made, except –

  (i) where before that person is called the court allows evidence of the making of the statement to be given on behalf of that party by some other person; or

  (ii) in so far as the court allows the person by whom the statement was made to narrate it in the course of his examination-in-chief on the ground that to prevent him from doing so would adversely affect the intelligibility of his evidence.

(3) Where in any civil proceedings a statement which was made otherwise than in a document is admissible by virtue of this section, no evidence other than direct oral evidence by the person who made the statement or any person who heard or otherwise perceived it being made shall be admissible for the purpose of proving it:

Provided that if the statement in question was made by a person while giving oral evidence in some other legal proceedings (whether civil or criminal), it may be proved in any manner authorised by the court.

Witness's previous statement, if proved, to be evidence of facts stated

3. – (1) Where in any civil proceedings –

(*a*) a previous inconsistent or contradictory statement made by a person called as a witness in those proceedings is proved by virtue of section 3, 4 or 5 of the Criminal Procedure Act 1865; or

(*b*) a previous statement made by a person called as aforesaid is proved for the purpose of rebutting a suggestion that his evidence has been fabricated,

that statement shall by virtue of this subsection be admissible as evidence of any fact stated therein of which direct oral evidence by him would be admissible.

(2) Nothing in this Act shall affect any of the rules of law relating to the circumstances in which, where a person called as a witness to any civil proceedings is cross-examined on a document used by him to refresh his

memory, that document may be made evidence in those proceedings; and where a document or any part of a document is received in evidence in any such proceedings by virtue of any such rule of law, any statement made in that document or part by the person using the document to refresh his memory shall by virtue of this subsection be admissible as evidence of any fact stated therein of which direct oral evidence by him would be admissible.

Admissibility of certain records as evidence of facts stated

4. – (1) Without prejudice to section 5 of this Act, in any civil proceedings a statement contained in a document shall, subject to this section and to rules of court, be admissible as evidence of any fact stated therein of which direct oral evidence would be admissible, if the document is, or forms part of, a record compiled by a person acting under a duty from information which was supplied by a person (whether acting under a duty or not) who had, or may reasonably be supposed to have had, personal knowledge of the matters dealt with in that information and which, if not supplied by that person to the compiler of the record directly, was supplied by him to the compiler of the record indirectly through one or more intermediaries each acting under a duty.

(2) Where in any civil proceedings a party desiring to give a statement in evidence by virtue of this section has called or intends to call as a witness in the proceedings the person who originally supplied the information from which the record containing the statement was compiled, the statement –

(*a*) shall not be given in evidence by virtue of this section on behalf of that party without the leave of the court; and

(*b*) without prejudice to paragraph (*a*) above, shall not without the leave of the court be given in evidence by virtue of this section on behalf of that party before the conclusion of the examination-in-chief of the person who originally supplied the said information.

(3) Any reference in this section to a person acting under a duty includes a reference to a person acting in the course of any trade, business, profession or other occupation in which he is engaged or employed or for the purposes of any paid or unpaid office held by him.

Admissibility of statements produced by computers

5. – (1) In any civil proceedings a statement contained in a document produced by a computer shall, subject to rules of court, be admissible as evidence of any fact stated therein of which direct oral evidence would be admissible, if it is shown that the conditions mentioned in subsection (2) below are satisfied in relation to the statement and computer in question.

(2) The said conditions are –

(a) that the document containing the statement was produced by the computer during a period over which the computer was used regularly to store or process information for the purposes of any activities regularly carried on over that period, whether for profit or not, by any body, whether corporate or not, or by any individual;

(b) that over that period there was regularly supplied to the computer in the ordinary course of those activities information of the kind contained in the statement or of the kind from which the information so contained is derived;

(c) that throughout the material part of that period the computer was operating properly or, if not, that any respect in which it was not operating properly or was out of operation during that part of that period was not such as to affect the production of the document or the accuracy of its contents; and

(d) that the information contained in the statement reproduces or is derived from information supplied to the computer in the ordinary course of those activities.

(3) Where over a period the function of storing or processing information for the purposes of any activities regularly carried on over that period as mentioned in subsection 2(a) above was regularly performed by computers, whether –

(a) by a combination of computers operating over that period; or

(b) by different computers operating in succession over that period; or

(c) by different combinations of computers operating in succession over that period; or

(d) in any other manner involving the successive operation over that period, in whatever order, of one or more computers and one or more combinations of computers,

all the computers used for that purpose during that period shall be treated for the purposes of this Part of this Act as constituting a single computer; and references in this Part of this Act to a computer shall be construed accordingly.

(4) In any civil proceedings where it is desired to give a statement in evidence by virtue of this section, a certificate doing any of the following things, that is to say –

(a) identifying the document containing the statement and describing the manner in which it was produced;

(b) giving such particulars of any device involved in the production of that document as may be appropriate for the purpose of showing that the document was produced by a computer;

(c) dealing with any of the matters to which the conditions mentioned in subsection (2) above relate,

and purporting to be signed by a person occupying a responsible position in relation to the operation of the relevant device or the management of the relevant activities (whichever is appropriate) shall be evidence of any matter stated in the certificate; and for the purposes of this subsection it shall be sufficient for a matter to be stated to the best of the knowledge and belief of the person stating it.

(5) For the purposes of this Part of this Act –

(a) information shall be taken to be supplied to a computer if it is supplied thereto in any appropriate form and whether it is so supplied directly or (with or without human intervention) by means of any appropriate equipment;

(b) where, in the course of activities carried on by any individual or body, information is supplied with a view to its being stored or processed for the purposes of those activities by a computer operated otherwise than in the course of those activities, that information, if duly supplied to that computer, shall be taken to be supplied to it in the course of those activities;

(c) a document shall be taken to have been produced by a computer whether it was produced by it directly or (with or without human intervention) by means of any appropriate equipment.

(6) Subject to subsection (3) above, in this Part of this Act 'computer' means any device for storing and processing information, and any reference to information being derived from other information is a reference to its being derived therefrom by calculation, comparison or any other process.

Provisions supplementary to sections 2 to 5

6. – (1) Where in any civil proceedings a statement contained in a document is proposed to be given in evidence by virtue of section 2, 4 or 5 of this Act it may, subject to any rules of court, be proved by the production of that document or (whether or not that document is still in existence) by the production of a copy of that document, or of the material part thereof, authenticated in such manner as the court may approve.

(2) For the purpose of deciding whether or not a statement is admissible in evidence by virtue of section 2, 4 or 5 of this Act, the court may draw

any reasonable inference from the circumstances in which the statement was made or otherwise came into being or from any other circumstances, including, in the case of a statement contained in a document, the form and contents of that document.

(3) In estimating the weight, if any, to be attached to a statement admissible in evidence by virtue of section 2, 3, 4 or 5 of this Act regard shall be had to all the circumstances from which any inference can reasonably be drawn as to the accuracy or otherwise of the statement and, in particular –

> (*a*) in the case of a statement falling within section 2(1) or 3(1) or (2) of this Act, to the question whether or not the statement was made contemporaneously with the occurrence or existence of the facts stated, and to the question whether or not the maker of the statement had any incentive to conceal or misrepresent the facts;
>
> (*b*) in the case of a statement falling within section 4(1) of this Act, to the question whether or not the person who originally supplied the information from which the record containing the statement was compiled did so contemporaneously with the occurrence or existence of the facts dealt with in that information, and to the question whether or not that person, or any person concerned with compiling or keeping the record containing the statement, had any incentive to conceal or misrepresent the facts; and
>
> (*c*) in the case of a statement falling within section 5(1) of this Act, to the question whether or not the information which the information contained in the statement reproduces or is derived from was supplied to the relevant computer, or recorded for the purpose of being supplied thereto, contemporaneously with the occurrence or existence of the facts dealt with in that information, and to the question whether or not any person concerned with the supply of information to that computer, or with the operation of that computer or any equipment by means of which the document containing the statement was produced by it, had any incentive to conceal or misrepresent the facts.

(4) For the purpose of any enactment or rule of law or practice requiring evidence to be corroborated or regulating the manner in which uncorroborated evidence is to be treated –

> (*a*) a statement which is admissible in evidence by virtue of section 2 or 3 of this Act shall not be capable of corroborating evidence given by the maker of the statement; and
>
> (*b*) a statement which is admissible in evidence by virtue of section 4 of this Act shall not be capable of corroborating evidence given by the person who originally supplied the information from which the record containing the statement was compiled.

(5) If any person in a certificate tendered in evidence in civil proceedings by virtue of section 5(4) of this Act wilfully makes a statement material in those proceedings which he knows to be false or does not believe to be true, he shall be liable on conviction on indictment to imprisonment for a term not exceeding two years or a fine or both.

Admissibility of evidence as to credibility of maker etc. of statement admitted under 2 or 4

7.—(1) Subject to rules of court, where in any civil proceedings a statement made by a person who is not called as a witness in those proceedings is given in evidence by virtue of section 2 of this Act—

(*a*) any evidence which, if that person had been so called, would be admissible for the purpose of destroying or supporting his credibility as a witness shall be admissible for that purpose in those proceedings; and

(*b*) evidence tending to prove that, whether before or after he made that statement, that person made (whether orally or in a document or otherwise) another statement inconsistent therewith shall be admissible for the purpose of showing that that person has contradicted himself:

Provided that nothing in this subsection shall enable evidence to be given of any matter of which, if the person in question had been called as a witness and had denied that matter in cross-examination, evidence could not have been adduced by the cross-examining party.

(2) Subsection (1) above shall apply in relation to a statement given in evidence by virtue of section 4 of this Act as it applies in relation to a statement given in evidence by virtue of section 2 of this Act, except that references to the person who made the statement and to his making the statement shall be construed respectively as references to the person who originally supplied the information from which the record containing the statement was compiled and to his supplying that information.

(3) Section 3(1) of this Act shall apply to any statement proved by virtue of subsection (1)(*b*) above as it applies to a previous inconsistent or contradictory statement made by a person called as a witness which is proved as mentioned in paragraph (*a*) of the said section 3(1).

Rules of court

8.—(1) Provision shall be made by rules of court as to the procedure which, subject to any exceptions provided for in the rules, must be followed and the other conditions which, subject as aforesaid, must be fulfilled before a statement can be given in evidence in civil proceedings by virtue of section 2, 4 or 5 of this Act.

(2) Rules of court made in pursuance of subsection (1) above shall in particular, subject to such exceptions (if any) as may be provided for in the rules—

(*a*) require a party to any civil proceedings who desires to give in evidence any such statement as is mentioned in that subsection to give to every other party to the proceedings such notice of his desire to do so and such particulars of or relating to the statement as may be specified in the rules, including particulars of such one or more of the persons connected with the making or recording of the statement or, in the case of a statement falling within section 5(1) of this Act, such one or more of the persons concerned as mentioned in section 6(3)(*c*) of this Act as the rules may in any case require; and

(*b*) enable any party who receives such notice as aforesaid by counternotice to require any person of whom particulars were given with the notice to be called as a witness in the proceedings unless that person is dead, or beyond the seas, or unfit by reason of his bodily or mental condition to attend as a witness, or cannot with reasonable diligence be identified or found, or cannot reasonably be expected (having regard to the time which has elapsed since he was connected or concerned as aforesaid and to all the circumstances) to have any recollection of matters relevant to the accuracy or otherwise of the statement.

(3) Rules of court made in pursuance of subsection (1) above—

(*a*) may confer on the court in any civil proceedings a discretion to allow a statement falling within section 2(1), 4(1) or 5(1) of this Act to be given in evidence notwithstanding that any requirement of the rules affecting the admissibility of that statement has not been complied with, but except in pursuance of paragraph (*b*) below shall not confer on the court a discretion to exclude such a statement where the requirements of the rules affecting its admissibility have been complied with;

(*b*) may confer on the court power, where a party to any civil proceedings has given notice that he desires to give in evidence—

(i) a statement falling within section 2(1) of this Act which was made by a person, whether orally or in a document, in the course of giving evidence in some other legal proceedings (whether civil or criminal); or

(ii) a statement falling within section 4(1) of this Act which is contained in a record of any direct oral evidence given in some other legal proceedings (whether civil or criminal), to give directions on the application of any party to the proceedings as to whether, and if so on what conditions, the party desiring to give the statement in evidence will be permitted to do so and (where applicable) as to the manner in which that statement and any other evidence given in those other proceedings is to be proved; and

(c) may make different provision for different circumstances, and in particular may make different provision with respect to statements falling within sections 2(1), 4(1) and 5(1) of this Act respectively;

and any discretion conferred on the court by rules of court made as aforesaid may be either a general discretion or a discretion exercisable only in such circumstances as may be specified in the rules.

(4) Rules of court may make provision for preventing a party to any civil proceedings (subject to any exceptions provided for in the rules) from adducing in relation to a person who is not called as a witness in those proceedings any evidence which could otherwise be adduced by him by virtue of section 7 of this Act unless that party has in pursuance of the rules given in respect of that person such a counter-notice as is mentioned in subsection (2)(*b*) above.

(5) In deciding for the purposes of any rules of court made in pursuance of this section whether or not a person is fit to attend as a witness, a court may act on a certificate purporting to be a certificate of a fully registered medical practitioner.

(6) Nothing in the foregoing provisions of this section shall prejudice the generality of section 99 of the Supreme Court of Judicature (Consolidation) Act 1925, section 102 of the County Courts Act 1959, section 15 of the Justices of the Peace Act 1949 or any other enactment conferring power to make rules of court; and nothing in section 101 of the Supreme Court of Judicature (Consolidation) Act 1925, section 102(2) of the County Courts Act 1959 or any other enactment restricting the matters with respect to which rules of court may be made shall prejudice the making of rules of court with respect to any matter mentioned in the foregoing provisions of this section or the operation of any rules of court made with respect to any such matter.

Admissibility of certain hearsay evidence formerly admissible at common law

9.—(1) In any civil proceedings a statement which, if this Part of this Act had not been passed, would by virtue of any rule of law mentioned in subsection (2) below have been admissible as evidence of any fact stated therein shall be admissible as evidence of that fact by virtue of this subsection.

(2) The rules of law referred to in subsection (1) above are the following, that is to say any rule of law—

(a) whereby in any civil proceedings an admission adverse to a party to the proceedings, whether made by that party or by another person, may be given in evidence against that party for the purpose of proving any fact stated in the admission;

(b) whereby in any civil proceedings published works dealing with matters of a public nature (for example, histories, scientific works, dictionaries and maps) are admissible as evidence of facts of a public nature stated therein;

(c) whereby in any civil proceedings public documents (for example, public registers, and returns made under public authority with respect to matters of public interest) are admissible as evidence of facts stated therein; or

(d) whereby in any civil proceedings records (for example, the records of certain courts, treaties, Crown grants, pardons and commissions) are admissible as evidence of facts stated therein.

In this subsection 'admission' includes any representation of fact, whether made in words or otherwise.

(3) In any civil proceedings a statement which tends to establish reputation or family tradition with respect to any matter and which, if this Act had not been passed, would have been admissible in evidence by virtue of any rule of law mentioned in subsection (4) below—

(a) shall be admissible in evidence by virtue of this paragraph in so far as it is not capable of being rendered admissible under section 2 or 4 of this Act; and

(b) if given in evidence under this Part of this Act (whether by virtue of paragraph (a) above or otherwise) shall by virtue of this paragraph be admissible as evidence of the matter reputed or handed down;

and, without prejudice to paragraph (b) above, reputation shall for the purposes of this Part of this Act be treated as a fact and not as a statement or multiplicity of statements dealing with matter the reputed.

(4) The rules of law referred to in subsection (3) above are the following, that is to say any rule of law—

(a) whereby in any civil proceedings evidence of a person's reputation is admissible for the purpose of establishing his good or bad character;

(b) whereby in any civil proceedings involving a question of pedigree or in which the existence of a marriage is in issue evidence of reputation or family tradition is admissible for the purpose of proving or disproving pedigree or the existence of the marriage, as the case may be; or

(c) whereby in any civil proceedings evidence of reputation or family tradition is admissible for the purpose of proving or disproving the existence of any public or general right or of identifying any person or thing.

(5) It is hereby declared that in so far as any statement is admissible in any civil proceedings by virtue of subsection (1) or (3)(a) above, it may be given in evidence in those proceedings notwithstanding anything in sections 2 to 7 of this Act or in any rules of court made in pursuance of section 8 of this Act.

(6) The words in which any rule of law mentioned in subsection (2) or (4) above is there described are intended only to identify the rule in question and shall not be construed as altering that rule in any way.

Interpretation of Part I, and application to arbitrations, etc.

10.—(1) In this Part of this Act—
'computer' has the meaning assigned by section 5 of this Act;
'document' includes, in addition to a document in writing—

(a) any map, plan, graph or drawing;

(b) any photograph;

(c) any disc, tape, sound track or other device in which sounds or other data (not being visual images) are embodied so as to be capable (with or without the aid of some other equipment) of being reproduced therefrom; and

(d) any film, negative, tape or other device in which one or more visual images are embodied so as to be capable (as aforesaid) of being reproduced therefrom;

'film' includes a microfilm;

'statement' includes any representation of fact, whether made in words or otherwise.

(2) In this Part of this Act any reference to a copy of a document includes—

(a) in the case of a document falling within paragraph (c) but not (d) of the definition of 'document' in the foregoing subsection, a transcript of the sounds or other data embodied therein;

(b) in the case of a document falling within paragraph (*d*) but not (*c*) of that definition, a reproduction or still reproduction of the image or images embodied therein, whether enlarged or not;

(c) in the case of a document falling within both those paragraphs, such a transcript together with such a still reproduction; and

(d) in the case of a document not falling within the said paragraph (*d*) of which a visual image is embodied in a document falling within that paragraph, a reproduction of that image, whether enlarged or not,

and any reference to a copy of the material part of a document shall be construed accordingly.

(3) For the purposes of the application of this Part of this Act in relation to any such civil proceedings as are mentioned in section 18(1)(*a*) and (*b*) of this Act, any rules of court made for the purposes of this Act under section 99 of the Supreme Court of Judicature (Consolidation) Act 1925 shall (except in so far as their operation is excluded by agreement) apply, subject to such modifications as may be appropriate, in like manner as they apply in relation to civil proceedings in the High Court:

Provided that in the case of a reference under section 92 of the County Courts Act 1959 this subsection shall have effect as if for the references to the said section 99 and to civil proceedings in the High Court there were substituted respectively references to section 102 of the County Courts Act 1959 and to proceedings in a county court.

(4) If any question arises as to what are, for the purposes of any such civil proceedings as are mentioned in section 18(1)(*a*) or (*b*) of this Act, the appropriate modifications of any such rule of court as is mentioned in subsection (3) above, that question shall, in default of agreement, be determined by the tribunal or the arbitrator or umpire, as the case may be.

. . . . . .

Privilege

Privilege against incrimination of self or spouse
14.—(1) The right of a person in any legal proceedings other than criminal proceedings to refuse to answer any question or produce any document or thing if to do so would tend to expose that person to proceedings for an offence or for the recovery of a penalty—

(a) shall apply only as regards criminal offences under the law of any part of the United Kingdom and penalties provided for by such law; and (*b*) shall include a like right to refuse to answer any question or produce any document or thing if to do so would tend to expose the husband or wife of that person to proceedings for any such criminal offence or for the recovery of any such penalty.

(2) In so far as any existing enactment conferring (in whatever words) powers of inspection or investigation confers on a person (in whatever words) any right otherwise than in criminal proceedings to refuse to answer any question or give any evidence tending to incriminate that person, subsection (1) above shall apply to that right as it applies to the right described in that subsection; and every such existing enactment shall be construed accordingly.

(3) In so far as any existing enactment provides (in whatever words) that any proceedings other than criminal proceedings a person shall not be excused from answering any question or giving any evidence on the ground that to do so may incriminate that person, that enactment shall be construed as providing also that in such proceedings a person shall not be excused from answering any question or giving any evidence on the ground that to do so may incriminate the husband or wife of that person.

(4) Where any existing enactment (however worded) that—

(*a*) confers powers of inspection or investigation; or

(*b*) provides as mentioned in subsection (3) above,

further provides (in whatever words) that any answer or evidence given by a person shall not be admissible in evidence against that person in any proceedings or class of proceedings (however described, and whether criminal or not), that enactment shall be construed as providing also that any answer or evidence given by that person shall not be admissible in evidence against the husband or wife of that person in the proceedings or class of proceedings in question.

(5) In this section 'existing enactment' means any enactment passed before this Act; and the references to giving evidence are references to giving evidence in any manner, whether by furnishing information, making discovery, producing documents or otherwise.

. . . . . .

Abolition of certain privileges
16.—(1) The following rules of law are hereby abrogated except in relation to criminal proceedings, that is to say—

(*a*) the rule whereby, in any legal proceedings, a person cannot be compelled to answer any question or produce any document or thing if to do so would tend to expose him to a forfeiture; and (*b*) the rule whereby, in any legal proceedings, a person other than a party to the proceedings cannot be compelled to produce any deed or other document relating to his title to any land.

(2) The rule of law whereby, in any civil proceedings, a party to the proceedings cannot be compelled to produce any document relating solely to his own case and in no way tending to impeach that case or support the case of any opposing party is hereby abrogated.

(3) Section 3 of the Evidence (Amendment) Act 1853 (which provides that a husband or wife shall not be compellable to disclose any communication made to him or her by his or her spouse during the marriage) shall cease to have effect except in relation to criminal proceedings.

(4) In section 43(1) of the Matrimonial Causes Act 1965 (under which the evidence of a husband or wife is admissible in any proceedings to prove that marital intercourse did or did not take place between them during any period, but a husband or wife is not compellable in any proceedings to give evidence of the matters aforesaid), the words from 'but a husband or wife' to the end of the subsection shall cease to have effect except in relation to criminal proceedings.

(5) A witness in any proceedings instituted in consequence of adultery, whether a party to the proceedings or not, shall not be excused from answering any question by reason that it tends to show that he or she has been guilty of adultery; and accordingly the proviso to section 3 of the Evidence Further Amendment Act 1869 and, in section 43(2) of the Matrimonial Causes Act 1965, the words from 'but' to the end of the subsection shall cease to have effect.

. . . .

General

General interpretation and savings
18.—(1) In this Act 'civil proceedings' includes, in addition to civil proceedings in any of the ordinary courts of law—

(*a*) civil proceedings before any other tribunal, being proceedings in relation to which the strict rules of evidence apply; and

(*b*) an arbitration or reference, whether under an enactment or not,

but does not include civil proceedings in relation to which the strict rules of evidence do not apply.

(2) In this Act—

'court' does not include a court-martial, and, in relation to an arbitration or reference, means the arbitrator or umpire and, in relation to proceedings before a tribunal (not being one of the ordinary courts of law), means the tribunal;

'legal proceedings' includes an arbitration or reference, whether under an enactment or not;

and for the avoidance of doubt it is hereby declared that in this Act, and in any amendment made by this Act in any other enactment, references to a person's husband or wife do not include references to a person who is no longer married to that person.

(3) Any reference in this Act to any other enactment is a reference thereto as amended, and includes a reference thereto as applied, by or under any other enactment.

(4) Nothing in this Act shall prejudice the operation of any enactment which provides (in whatever words) that any answer or evidence given by a person in specified circumstances shall not be admissible in evidence against him or some other person in any proceedings or class of proceedings (however described).

In this subsection the reference to giving evidence is a reference to giving evidence in any manner, whether by furnishing information, making discovery, producing documents or otherwise.

(5) Nothing in this Act shall prejudice—

(*a*) any power of a court, in any legal proceedings, to exclude evidence (whether by preventing questions from being put or otherwise) at its discretion; or

(*b*) the operation of any agreement (whenever made) between the parties to any legal proceedings as to the evidence which is to be admissible (whether generally or for any particular purpose) in those proceedings.

(6) It is hereby declared that where, by reason of any defect of speech or hearing from which he is suffering, a person called as a witness in any legal proceedings gives his evidence in writing or by signs, that evidence is to be treated for the purposes of this Act as being given orally.

. . . . . .

Short title, repeals, extent and commencement

20.—(1) This Act may be cited as the Civil Evidence Act 1968.

(2) Sections 1, 2, 6(1) (except the words from 'Proceedings' to 'references') and 6(2)(*b*) of the Evidence Act 1938 are hereby repealed.

(3) This Act shall not extend to Scotland or, except in so far as it enlarges the powers of the Parliament of Northern Ireland, to Northern Ireland.

(4) The following provisions of this Act, namely sections 13 to 19, this section (except subsection (2)) and the Schedule, shall come into force on the day this Act is passed, and the other provisions of this Act shall come into force on such day as the Lord Chancellor may by order made by statutory instrument appoint; and different days may be so appointed for different purposes of this Act or for the same purposes in relation to different courts or proceedings or otherwise in relation to different circumstances.

ARBITRATION (COMMODITY CONTRACTS) ORDER 1979

(SI 1979, No. 754)

(Made under s4 of the 1979 Act).

1.—(1) This Order may be cited as the Arbitration (Commodity Contracts) Order 1979 and shall come into operation on 1st August 1979.

(2) In this Order—

'the Act' means the Arbitration Act 1979.

'market' means a commodity market or exchange.

2. The following markets are hereby specified for the purpose of section 4 of the Act—

(*a*) the markets set out in Part I of the Schedule hereto:

(*b*) any market in which contracts for sale are subject to the rules or regulations of one or other of the associations set out in Part II of the Schedule, whether or not the market is a market on which commodities are bought and sold at a particular place.

3. The following descriptions of contract are hereby specified for the purpose of section 4 of the Act—

(*a*) contracts for the sale of goods on any market specified in Article 2 of this Order;

(b) contracts for the sale of goods which are subject to arbitration rules of the London Metal Exchange or of an association set out in Part II of the Schedule hereto.

SCHEDULE

PART I

Markets

The London Cocoa Terminal Market
The London Coffee Terminal Market
The London Grain Futures Market
The London Metal Exchange
The London Rubber Terminal Market
The Gafta Soya Bean Meal Futures Market
The London Sugar Terminal Market
The London Vegetable Oil Terminal Market
The London Wool Terminal Market

PART II

Markets in which contracts are subject to rules or regulations of the following Associations
The Cocoa Association of London Limited
The Coffee Trade Federation
The Combined Edible Nut Trade Association
Federation of Oils, Seeds and Fats Associations Limited
The General Produce Brokers' Association of London
The Grain and Feed Trade Association Limited
The Hull Seed, Oil and Cake Association
The Liverpool Cotton Association Limited
London Jute Association
London Rice Brokers' Association
The National Federation of Fruit and Potato Traders Limited
The Rubber Trade Association of London
Skin, Hide and Leather Traders' Association Limited
The Sugar Association of London
The Refined Sugar Association
The Tea Brokers' Association of London
The British Wool Confederation

RULES OF THE SUPREME COURT
ORDER 73

ARBITRATION PROCEEDINGS

Arbitration proceedings not to be assigned to Chancery Division (Ord. 73, r. 1)

1.—(1) [*Revoked by RSC (Amendment No. 2) 1983 (S.I. 1983 No. 1181).*]

Matters for a judge in court (Ord. 73, r.2)

2.—Every application to the Court—

(*a*) to remit an award under section 22 of the Arbitration Act 1950, or

(*b*) to remove an arbitrator or umpire under section 23(1) of that Act, or

(*c*) to set aside an award under section 23(2) thereof, or

(*d*) [*Revoked by RSC (Amendment No. 2) 1983 (S.I. 1983 No. 1181).*]

(*e*) to determine, under section 2(1) of that Act, any question of law arising in the course of a reference,

must be made by originating motion to a single judge in court.

(2) Any appeal to the High Court under section 1(2) of the Arbitration Act 1979 shall be made by originating motion to a single judge in court and notice thereof may be included in the notice of application for leave to appeal, where leave is required.

(3) An application for a declaration that an award made by an arbitrator or umpire is not binding on a party to the award on the ground that it was made without jurisdiction may be made by originating motion to a single judge in court, but the foregoing provision shall not be taken as affecting the judge's power to refuse to make such a declaration in proceedings begun by motion.

Amended by RSC (Amendment No. 3) 1979 (S.I. 1979 No. 522) and RSC (Amendment No. 3) (S.I. 1983 No. 1181);

Matters for judge in chambers or master (Ord. 73, r. 3)

3.—(1) Subject to the foregoing provisions of this Order and the provisions of this rule, the jurisdiction of the High Court or a judge thereof under the Arbitration Act 1950 and the jurisdiction of the High Court under the Arbitration Act 1975 and the Arbitration Act 1979 may be exercised by a judge in chambers, a master or the Admiralty Registrar.

(2) Any application

(a) for leave to appeal under s. 1(2) of the Arbitration Act 1979, or

(b) under s 1(5) of that Act (including any application for leave), or

(c) under s 5 of that Act.

shall be made to a judge in chambers.

(3) Any application to which this rule applies shall, where an action is pending, be made by summons in the action, and in any other case by an originating summons which shall be in Form No. 10 in Appendix A.

(4) Where an application is made under section 1(5) of the Arbitration Act 1979 (including any application for leave) the summons must be served on the arbitrator or umpire and on any other party to the reference.

Amended by RSC (Amendment No. 3) 1979 (S.I. 1979 No. 522); RSC (Writ and Appearance) 1979 (S.I. 1979 No. 1716) and RSC (Amendment No. 2) 1983 (S.I. 1983 No. 1181).

Applications in district registries (Ord. 73, r. 4)

4. An application under section 12(4) or the Arbitration Act 1950 for an order that a writ of subpoena ad testificandum or of subpoena duces tecum shall issue to compel the attendance before an arbitrator or umpire of a witness may, if the attendance of the witness is required within the district of any district registry, be made at that registry, instead of at the Central Office, at the option of the applicant.

This Rule was taken from RSC (Rev.) 1962, [Ord. 88, r. 4, which had been taken from the former O. 37, r. 27B.

Time-limits and other special provisions as to appeals and applications under the Arbitration Acts (Ord. 73, r. 5)

5.—(1) An application to the Court—

(a) to remit an award under section 22 of the Arbitration Act 1950, or

(b) to set aside an award under section 23(2) of that Act or otherwise, or

(c) to direct an arbitrator or umpire to state the reasons for an award under section 1(5) of the Arbitration Act 1979,

must be made, and the summons or notice must be served, within 21 days after the award has been made and published to the parties.

(2) In the case of an appeal to the Court under section 1(2) of the Arbitration Act 1979, the Summons for leave to appeal, where leave is required, and the notice of originating motion must be served and the appeal entered within 21 days after the award has been made and published to the parties:

Provided that, where reasons material to the appeal are given on a date subsequent to the publication of the award, the period of 21 days shall run from the date on which the reasons are given.

(3) An application, under section 2(1) of the Arbitration Act 1979, to determine any question of law arising in the course of a reference, must be made, and notice thereof served, within 14 days after the arbitrator or umpire has consented to the application being made, or the other parties have so consented.

(4) For the purpose of paragraph (2) the consent must be given in writing.

(5) In the case of every appeal or application to which this rule applies, the notice of originating motion or, as the case may be, the originating summons, must state the grounds of the appeal of application and, where the appeal or application is founded on evidence by affidavit, or is made with the consent of the arbitrator or umpire or of the other parties, a copy of every affidavit intended to be used, or, as the case may be, of every consent given in writing, must be served with that notice.

Substituted by RSC (Amendment No. 3) 1979 (S.I. 1979 No. 522).
Amended by RSC (Amendment) 1986.

Applications and appeals to be heard by Commercial Judges

6.—(1) Any matter which is required, by rule 2 or 3, to be heard by a judge, shall be heard by a Commercial Judge, unless any such judge otherwise directs.

(2) Nothing in the foregoing paragraph shall be construed as preventing the powers of a Commercial Judge from being exercised by any judge of the High Court.

Substituted by RSC (Amendment No. 3) 1979 (S.I. 1979 No. 522).

Service out of the jurisdiction of summons, notice, etc. (Ord. 73, r. 7)

7.—(1) Service out of the jurisdiction—

(a) of an originating summons for the appointment of an arbitrator or umpire, or

(b) of notice of an originating motion to remove an arbitrator or umpire or to remit or set aside an award, or

(c) of an originating summons or notice of an originating motion under the Arbitration Act 1979, or

(d) of any order made on such a summons or motion as aforesaid, is permissible with the leave of the Court provided that the arbitration to which the summons, motion or order relates is governed by English law or has been, is being, or is to be held, within the jurisdiction.

(1A) Service out of the jurisdiction of an originating summons for leave to enforce an award is permissible with the leave of the Court whether or not the arbitration is governed by English law.

(2) An application for the grant of leave under this rule must be supported by an affidavit stating the grounds on which the application is made and showing in what place or country the person to be served is, or probably may be found; and no such leave shall be granted unless it shall be made sufficiently to appear to the Court that the case is a proper one for service out of the jurisdiction under this rule.

(3) Order 11, rules 5, 6 and 8, shall apply in relation to any such summons, notice or order as is referred to in paragraph (1) as they apply in relation to a writ.

Amended by RSC (Amendment No. 4) 1979 (S.I. 1979 No. 1542); RSC (Amendment No. 4) 1980 (S.I. 1980 No. 2000) and (with effect from the date s2 of the Civil Jurisdiction and Judgments Act 1982 comes into force) by RSC (Amendment No. 2) 1983 (S.I. 1983 No. 1181).

Registration in High Court of foreign awards (Ord. 73, r. 8)

8. Where an award is made in proceedings on an arbitration in any part of Her Majesty's dominions or other territory to which Part I of the Foreign Judgments (Reciprocal Enforcement) Act 1933 extends, being a part to which Part II of the Administration of Justice Act 1920 extended immediately before the said Part I was extended thereto, then, if the award has, in pursuance of the law in force in the place where it was made, become enforceable in the same manner as a judgment given by a court in that place, Order 71 shall apply in relation to the award as it applies in relation to a judgment given by that court, subject, however, to the following modifications:—

(*a*) for references to the country of the original court there shall be substituted references to the place where the award was made; and

(*b*) the affidavit required by rule 3 of the said Order must state (in addition to the other matters required by that rule) that to the best of the information or belief of the deponent the award has, in pursuance of the law in force in the place where it was made, become enforceable in the same manner as a judgment given by a court in that place.

Taken from RSC (Rev.) 1962, Ord. 88, r. 7, which replaced the former O. 41B, r. 15.

Registration of awards under Arbitration (International Investment Disputes) Act 1966 (Ord. 73, r. 9)

9.—(1) In this rule and in any provision of these rules as applied by this rule—

'the Act of 1966' means the Arbitration (International Investment Disputes) Act 1966;

'award' means an award rendered pursuant to the Convention;

'the Convention' means the convention referred to in section 1(1) of the Act of 1966;

'judgment creditor' and 'judgment debtor' means respectively the person seeking recognition or enforcement of an award and the other party to the award.

(2) Subject to the provisions of this rule, the following provisions of Order 71, namely, rules 1, 3(1) (except sub-paragraphs (c)(iv) and (d) thereof) 7 (except paragraph (3)(c) and (d) thereof), and 10(3) shall apply with the necessary modifications in relation to an award as they apply in relation to a judgment to which Part II of the Foreign Judgments (Reciprocal Enforcement) Act 1933 applies.

(3) An application to have an award registered in the High Court under section 1 of the Act of 1966 shall be made by originating summons which shall be in Form No. 10 in Appendix A.

(4) The affidavit required by Order 71, rule 3, in support of an application for registration shall—

(a) in lieu of exhibiting the judgment or a copy thereof, exhibit a copy of the award certified pursuant to the Convention, and

(b) in addition to stating the matters mentioned in paragraph 3(1)(c)(i) and (ii) of the said rule 3, state whether at the date of the application the enforcement of the award has been stayed (provisionally or otherwise) pursuant to the Convention and whether any, and if so what, application has been made pursuant to the Convention which, if granted, might result in a stay of the enforcement of the award.

(5) There shall be kept in the Central Office under the direction of the senior master a register of the awards ordered to be registered under the Act of 1966 and particulars shall be entered in the register of any execution issued on such an award.

(6) Where it appears to the court on granting leave to register an award or on an application made by the judgment debtor after an award has been registered—

(*a*) that the enforcement of the award has been stayed (whether provisionally or otherwise) pursuant to the Convention, or

(*b*) that an application has been made pursuant to the Convention which, if granted, might result in a stay of the enforcement of the award,

the court shall, or, in the case referred to in sub-paragraph (b) may, stay execution of the award for such time as it considers appropriate in the circumstances.

(7) An application by the judgment debtor under paragraph (6) shall be made by summons and supported by affidavit.

Added by RSC (Amendment No. 1) 1968 (S.I. 1968 No. 1244) and amended by RSC (Amendment No. 3) 1977 (S.I. 1977 No. 1955); RSC (Writ and Appearance) 1979 (S.I. 1979 No. 1716) and RSC (Amendment No. 2) 1982 (S.I. 1982 No. 1111).

Enforcement of arbitration awards (Ord. 73, r. 10)

10.—(1) An application for leave under section 26 of the Arbitration Act 1950 or under section 3(1)(*a*) of the Arbitration Act 1975 to enforce an award on an arbitration agreement in the same manner as a judgment or order may be made ex parte but the Court hearing the application may direct a summons to be issued.

(2) If the Court directs a summons to be issued, the summons shall be an originating summons which shall be in Form No. 10 in Appendix A.

(3) An application for leave must be supported by affidavit—

(*a*) exhibiting

(i) where the application is under section 26 of the Arbitration Act 1950, the arbitration agreement and the original award or, in either case, a copy thereof;

(ii) where the application is under section 3(1)(*a*) of the Arbitration Act 1975, the documents required to be produced by section 4 of that Act,

(*b*) stating the name and the usual or last known place of abode or business of the applicant (hereinafter referred to as 'the creditor') and the person against whom it is sought to enforce the award (herein after referred to as 'the debtor') respectively,

(*c*) as the case may require, either that the award has not been complied with or the extent to which it has not been complied with at the date of the application.

(4) An order giving leave must be drawn up by or on behalf of the creditor and must be served on the debtor by delivering a copy to him personally or by sending a copy to him at his usual or last known place of abode or business or in such other manner as the Court may direct.

(5) Service of the order out of the jurisdiction is permissible without leave, and Order 11, rules 5, 6 and 8, shall apply in relation to such an order as they apply in relation to a writ.

(6) Within 14 days after service of the order or, if the order is to be served out of the jurisdiction, within such other period as the Court may fix, the debtor may apply to set aside the order and the award shall not be enforced until after the expiration of that period or, if the debtor applies within that period to set aside the order, until after the application is finally disposed of.

(7) The copy of that order served on the debtor shall state the effect of paragraph (6).

(8) In relation to a body corporate this rule shall have effect as if for any reference to the place of abode or business of the creditor or the debtor there were substituted a reference to the registered or principal address of the body corporate; so, however, that nothing in this rule shall affect any enactment which provides for the manner in which a document may be served on a body corporate.

Added by RSC (Amendment No. 4) 1978 (S.I. 1978 No. 1066) and amended by RSC (Amendment) 1979 (S.I. 1979 No. 35); RSC (Writ and Appearance) 1979 (S.I. 1979 No. 1716) and RSC (Amendment No. 4) 1980 (S.I. 1980 No. 2000).

THE ARBITRATION (FOREIGN AWARDS) ORDER 1984
SI 1984/1168

(As amended by SI 1985/455, SI 1986/949 and SI 1987/1029)

GENEVA CONVENTION STATES

Column 1	Column 2
Powers party to the Geneva Convention	*Territories to which the Geneva Convention applies*
The United Kingdom of Great Britian and Northern Ireland	The United Kingdom of Great Britain and Northern Ireland
	Anguilla
	British Virgin Islands
	Cayman Islands
	Falkland Islands
	Falkland Islands Dependencies
	Gibraltar
	Hong Kong
	Montserrat
	Turks and Caicos Islands
Antigua and Barbuda	Antigua and Barbuda
Austria	Austria
Bahamas	Bahamas
Bangladesh	Bangladesh
Belgium	Belgium
Belize	Belize
Czechoslovakia	Czechoslovakia
Denmark	Denmark
Dominica	Dominica
Finland	Finland
France	France
Federal Republic of Germany	Federal Republic of Germany
German Democratic Republic	German Democratic Republic
Greece	Greece
Grenada	Grenada
Guyana	Guyana
India	India
Republic of Ireland	Republic of Ireland
Israel	Israel
Italy	Italy
Japan	Japan
Kenya	Kenya
Luxembourg	Luxembourg
Malta	Malta
Mauritius	Mauritius
Netherlands	Netherlands (including Curacao)
New Zealand	New Zealand
Pakistan	Pakistan
Portugal	Portugal
Romania	Romania
Saint Christopher and Nevis	Saint Christopher and Nevis
St. Lucia	St. Lucia
Spain	Spain
Sweden	Sweden
Switzerland	Switzerland

Column 1	Column 2
Powers party to the Geneva Convention	*Territories to which the Geneva Convention applies*
Tanzania	Tanzania
Thailand	Thailand
Western Samoa	Western Samoa
Yugoslavia	Yugoslavia
Zambia	Zambia

NEW YORK CONVENTION STATES

Australia (including all the external territories for the international relations of which Australia is responsible)	Japan
	Jordan
	Korea
Austria	Kuwait
Belgium	Luxembourg
Belize	Madagascar
Benin	Malaysia
Botswana	Mexico
Bulgaria	Monaco
Byelorussian Soviet Socialist Republic	Morocco
Cambodia	Netherland (including the Netherlands Antilles)
Central African Republic	New Zealand
Chile	Niger
China	Nigeria
Colombia	Norway
Cuba	Panama
Cyprus	Philippines
Czechoslovakia	Poland
Denmark (including Greenland and the Faroe Islands)	Romania
	San Marino
Djibouti	Singapore
Ecuador	South Africa
Egypt	Spain
Finland	Sri Lanka
France (including all the territories of the French Republic)	Sweden
	Switzerland
Federal Republic of Germany	Syria
German Democratic Republic	Tanzania
Ghana	Thailand
Greece	Trinidad and Tobago
Grenada	Tunisia
Guatemala	Ukrainian Soviet Socialist Republic
Haiti	Union of Soviet Socialist Republics
Holy See	United States of America (including all the territories for the international relations of which the United States of America is responsible)
Hungary	
India	
Indonesia	
Republic of Ireland	Uruguay
Israel	Yugoslavia
Italy	

INDEX

ABBREVIATIONS xxvii
AFFIDAVIT
  evidence by 43, 44
AMIABLE COMPOSITEUR 11, 13
ALTERNATIVE DISPUTE
    RESOLUTION 1, 70
APPEALS
  Arbitration Act 1979, under 92
  arbitration agreement, under 91
  Court of Appeal, to 95-6
  exclusion agreements 99
  judicial review 95
  leave to appeal 96-9
  preliminary point of law 93
  procedure 94
APPOINTMENT OF ARBITRATOR
  acceptance 38
  court, by the 41
  eligibility for, 6-7
  judicial 42
  parties, by the 28, 38-9
ARBITRATE
  oral agreement to, 27
ARBITRATION
  administered 50
  advantages and disadvantages of Ch 2
  attendance at 65
  claimant and respondent 52
  commencement of 7
  commodity 68
  consolidation 21, 28, 117
  construction industry 69
  convenience 19
  condition precendent to, 34
  costs 14
  definition 1
  documents only 64, 66
  form of Court procedure 62
  frustration 29-30
  legal aid 23
  litigation relationship to 8, 23
  look-sniff 64, 67
  matters suitable for reference to 4, 23
  multi-party dispute 20
  pendulum 68
  powers in 43-5
  publicity 15

  representation 18
  settlement of dispute during 60
  speed 16
  termination of before award 60
  time limitation 34-6
  what may be referred 4
  who may refer 5
  See also AWARD, ENFORCEMENT OF
    AWARD, HEARING,
    STATUTORY ARBITRATIONS,
    and STAY OF PROCEEDINGS
ARBITRATION AGREEMENT
  abandonment of 29, 37
  arrangement as to costs in 79, 80
  definition 25
  equity clauses 10
  enforcement 28
  form 25
  frustration of 29-30, 37
  independent existence as a contract 36
  mutuality 27
  order that it is to cease to have effect 34,
    36-7, 47-8
  proper law of 9
  repudiation of 29
  resisting enforcement of 29
  signature 26-7
  termination of 36-7
  terms implied 27-8
  torts, inclusion of 26
ARBITRATOR (ARBITRATORS)
  advocate acting as 6, 39, 78
  appointment of - See APPOINTMENT
  authority of, revoked 47
  bias of 6-7, 32, 46, 47, 91
  control of procedure 61-2
  disagreement of arbitrators 40
  discretion of as to costs 82
  disqualification of 6, 50
  duty of 72
  functus officio 73, 100
  jurisdiction of 4, 19, 43
  liability of 3
  lien of 86
  majority of 40, 78
  misconduct of 47, 81, 86-7, 88, 100-1
  number of 27, 40-1

powers 43, 44
   additional 44–5
qualifications of 6
reasons, to provide 93–4
removal of – See REMOVAL OF ARBITRATOR OR UMPIRE
remuneration of – See REMUNERATION
retirement 48
revocation of authority of 47
technical qualifications 17, 64
third 40
who may act 5
See INTERNATIONAL ARBITRATIONS, ARBITRATOR-ADVOCATES 6, 39, 78
ATLANTIC SHIPPING CLAUSE 8, 28 29, 34–6
   undue hardship 35
AWARD
   by consent 60
   clerical error in 73
   contents 72, 77
   convention 108–9
   costs of – See COSTS
   Court's attitude to 72, 91
   currency of 78
   declaratory 76, 87
   directions in 72, 76–7
   drawing up of, legal professional advice 66, 77
   effect of 8, 73
   enforcement of – See ENFORCEMENT OF AWARD
   finality 28, 73
   foreign 110–11
   form, of 76–8
   illustration 116
   interest 28
   interim 22, 27, 74, 80
   interim final 74
   joint execution of 78
   judgment merging in 87
   lien on 86
   majority of arbitrators, of 40, 78
   publication of 79
   reasoned 75–6
   remitter of – See REMISSION
   signature of 77, 78
   specific performance 27, 73
   taking up 86
   time for making 78–9
   vitality, its own 8
   See also REASONS, SETTING ASIDE

BIAS 6–7, 32, 46, 47

CALDERBANK OFFERS 84
CASE STATED abolished 92

CENTROCON CLAUSE – See ATLANTIC SHIPPING CLAUSE
CHARTERED INSTITUTE OF ARBITRATORS 5, 18, 23, 26, 52, 66, 81, 106
See also RULES OF PROCEDURE
CIVIL LAW
   discovery 55, 108
   evidence 107–8
COMMERCIAL COURT 64, 90–1
   practice note 17
COMMODITY ARBITRATIONS 68
CONCILIATION 1
CONSOLIDATION 21–2, 28, 117
CONSTRUCTION INDUSTRY ARBITRATIONS 69
CONVENTION AWARDS 108–9
COSTS
   agreement by parties as to 79–80
   award, of the 79
   award, should provide for 80
   discretion of arbitrator 28
   judicial exercise of discretion as to 82
   reference, of the 79
   sealed and without prejudice offers 83–4
   security for 44, 80
   shorthand note, of 56
   solicitor, of 15, 80
   successful party, against 82–3
   taxation of – See TAXATION OF COSTS
   VAT upon 84
COSTS ARBITRATION SERVICE 81
COURT OF APPEAL 43, 49, 95–8
COURT OF JUSTICE OF THE EEC 22, 98
COURTS
   attitude of, to arbitration 28
   control by Ch. 8
   inherent jurisdiction of 48, 102

DECLARATION 30, 60, 88
DIRECTIONS 57, 60, 112
DISCOVERY 44, 53, 54–5
   Crown, against 54
   privilege 54
DISPUTE
   existence of 33–4
DOCUMENTS
   agreed bundles 55
   discovery – See DISCOVERY
   inspection of 54–5
DOCUMENTS ONLY ARBITRATIONS 64, 66

ENFORCEMENT OF AWARD
   action, by 88
   defences 88
   foreign 18, 108–11

judgment, as a 87
EQUITY CLAUSES 10
EUROPEAN COURT OF JUSTICE 22, 98
EVIDENCE
 affidavit, by 44
 expert 65
 new 74, 101
 strict rules of, 10-11
 under civil law systems 107-8
 See also WITNESSES
EXCLUSION AGREEMENTS 48, 56, 92, 93, 95, 99-100
EX PARTE
 proceeding 58

FOREIGN AWARD 110-11
FRAUD 4, 36, 47-8, 65, 100, 101
FURTHER AND BETTER PARTICULARS 58

GENEVA CONVENTION 110
GUIDELINES
 the *Nema* 86-9
 undue hardship 35

HEARING Ch. 6
 attendance at 65
 preparations for 57
 provisional date for 57
HIGH TECHNOLOGY 70-1

INFANT 5-6
INJUNCTION 44, 47, 60, 102
INTEREST 85
 after award 85-6
INTERIM AWARD 22, 27, 74, 93
INTERNATIONAL ARBITRATIONS
 number of arbitrators 105
 rules of procedure 106-7
 terms of arbitration agreement 104-6
INTERROGATORIES 44, 59

JUDGE-ARBITRATORS 42-3, 45
JUDICIAL REVIEW 95

LAW
 applicable 9
LEGAL AID
 not available 23
 unavailability and grant of stay 33
LEGAL ASSESSORS 66
LEGAL ASSISTANCE 67
LIMITATION PERIOD 7-8

LONDON MARITIME ARBITRATORS ASSOCIATION See under RULES
 pre-trial procedure 59, 65
 See also RULES OF PROCEDURE
LOOK-SNIFF ARBITRATORS 64, 67

MEDIATION 1
MISCONDUCT
 arbitrator, of 47, 88, 100, 101
 excessive remuneration, in regard to 81, 86-7
 permitting intervention by the Courts 100 1

NEW EVIDENCE 74, 101
NEW YORK CONVENTION 1, 2, 18, 31, 34, 108-10
NON-COMPLIANCE WITH ARBITRATOR'S DIRECTIONS 44-5, 58
NOTICES TO ADMIT 58, 59

OATH 27
 examination of parties and witnesses 65
OFFERS 83-4
OFFICIAL REFEREE 17, 42

PARTY (PARTIES)
 Crown 5
 death of 36
 examination on oath 27, 43
 implied contract to abide by award 8
 recalcitrant 58
PENDULUM ARBITRATIONS 68
PEREMPTORY NOTICE 58, 115
PLEADINGS
 amendment of 59
 late summary of issues 59
 nature of 53
 Scott schedule 53, 114
 statements of case 53
PRACTICE NOTE
 Commercial Court 17
PRE-HEARING REVIEW 57, 59
PRELIMINARY MEETING 50-7
 LMAA cases 59
PRELIMINARY POINT OF LAW 93
PROCEDURE
 arbitrator's discretion 61-2
 flexibility 19
 High Court, of 62
 modification for arbitrations 63-4
 See RULES OF PROCEDURE
PROPERTY
 inspection of 44, 55
 preservation of 44, 56

QUALITY ARBITRATIONS – See COMMODITY ARBITRATIONS

REASONS 56-7, 93-4
  given separately 76
  time limits for applications for 94
REFEREES
  arbitral 69-70, 71
REMISSION (OR REFERRING BACK) 73-4, 79, 100-101
REMOVAL OF ARBITRATOR OR UMPIRE
  Court, by 7, 46, 47, 91
  deciding by reference to unargued points 18
  dilatory, for being 17, 46, 47
  effect of 42
  impartiality, for possibly lacking 47, 91
  misconduct, for 47
  parties by 46
REMUNERATION OF ARBITRATOR OR UMPIRE
  dilatory arbitrator 46
  excessive claims for 86, 87
  implied entitlement to charge 45-6
  lien in respect of 86
  quantum meruit basis 46
  recovery of, irrevocability of 46
  security for 45-6
  taxation of 81
RES JUDICATA 60
REVOCATION
  authority of arbitrator or umpire, of 47
RULES OF PROCEDURE 25, 43, 51-2
  arbitrator's 51-2
  Chartered Institute of Arbitrators 1, 26, 52
  Cocoa Association of London 91
  Grain and Feed Trade Association (GAFTA) 29, 52, 68, 69, 91
  Institution of Construction Engineers 69-70
  International Bar Association 107-8
  International Chamber of Commerce 106-7
  London Bar Arbitration Scheme 48
  London Court of International Arbitration 106
  London Maritime Arbitrators Association 17, 21, 39, 52, 59
  Refined Sugar Association 21, 41, 61
  Royal Institute of British Architects 52
  Uncitral 107

SCOTT V *AVERY* CLAUSES 8, 28, 34, 102
  courts discretion to override 35-6

SCOTT SCHEDULE 53, 70, 114
SEALED OFFERS 83-4
SECURITY FOR COSTS 32, 44, 45, 90, 101
SETTING ASIDE AWARD
  error of fact or law on its face, abolished 92
  misconduct of arbitrator 65, 73, 100-1
SHORTHAND WRITER 56, 65
SPECIAL CASE
  abolished 92
SPECIFIC PERFORMANCE 27, 28, 73, 88, 89
STATUTORY ARBITRATIONS
  enforcement 87
  exclusion agreement has no effect 13
  limitation period, and 7
  rules 12-13, 52
  Scott v *Avery* relieving provision does not apply 35
STAY OF PROCEEDINGS
  application for 28, 30
  dispute, existence of 30, 33-4
  divisible proceedings 31
  effect of 8
  merits, where applicant lacks 31
  readiness to do all things necessary, etc. 30
  refusal
    where arbitration agreement ordered to have no effect 31
    other grounds for 31
  statutory provisions 30
  step in proceedings, what constitutes 32
  sufficient reasons not to refer 30, 32
STRING ARBITRATIONS 68
SUMMARY OF ISSUES 59

TAXATION OF COSTS
  arbitrator's fees, taxation of 81
  Costs Arbitration Service 81-2
  High Court, by 81
  indemnity basis 81
  standard basis 81
  umpire, by 81
TIME
  award, for making 79
  enlargement of under s27 35-6
  limits for appeals 94
  applications to set aside or remit 94
TIME BAR CLAUSE
  discretion of court to extend time 35-6
  see *ATLANTIC SHIPPING* CLAUSE
TIME TABLE 52, 58

UMPIRE
  appointment 27, 39-40

entry on reference by 27, 39–40
judicial appointment 42–3
more than one 40
remuneration—see REMUNERATION
   OF ARBITRATOR
revocation of authority of 47

VAT 84
VALUATION
  arbitration, distinguished from 3

WANT OF PROSECUTION
  dismissal of case for 29, 45
WITNESSES
  commissions for the examination of 56
  evidence of in High Court 62, 67
  exchange of proofs by 70
  expert 65
    discussion with, before pleadings 51
    exchange and agreement of proofs
      by 56
  summoning 43–4